Commercial Plant Breeding
Volume 02: Field Crops

a division of
NIPA GENX ELECTRONIC RESOURCES & SOLUTIONS P. LTD.
New Delhi-110 034

Commercial Plant Breeding Volume 01 Vegetable Crops (Complete in Two Parts)

The present book entitled "Commercial Plant Breeding-1 Vegetable Crops" contains 32 chapters bifurcated into two parts, part I "Commercial Plant Breeding and Regulatory Affairs" having initial 14 chapters deals with commercial plant breeding, the vegetable seed business, vegetable seed supply chain management, all India coordinated research project (vegetable crops), world vegetable centre (AVRDC), and seed and IPR related acts. The part II "Commercial Breeding of Vegetable Crops" having remaining 18 chapters is devoted to commercial breeding of individual major vegetable crops relevant to seed industry.

The book is intended for undergraduate and post graduate students of state agricultural universities and agricultural colleges for the courses on vegetable breeding in general and commercial vegetable breeding in particular and the professionals across ICAR Institutes, SAUs and Seed Industry. First of its kind, the present book aims to bring academic and vegetable seed business worlds both on a common platform

Commercial Plant Breeding
Volume 2: Field Crops

Part 01: Commercial Plant Breeding Concepts and Regulatory Issues

---

Hari Har Ram

Former Professor
Department of Plant Breeding and Genetics
G.B. Pant University of Agriculture and Technology
Pantnagar, Udham Singh Nagar-263 145
Uttarakhand, India

Former Head
Department of Vegetable Science
G.B. Pant University of Agriculture and Technology
Pantnagar, Udham Singh Nagar-263 145
Uttarakhand, India

Former Vice President
Krishidhan Vegetable Seeds, Pune
Former FAO Consultant (Myanmar, Libya, DPR Korea)

a division of

NIPA GENX ELECTRONIC RESOURCES & SOLUTIONS P. LTD.

New Delhi-110 034

a division of

NIPA GENX ELECTRONIC RESOURCES & SOLUTIONS P. LTD.

101,103, Vikas Surya Plaza, CU Block
L.S.C.Market, Pitam Pura, New Delhi-110 034
Ph : +91 11 27341616, 27341717, 27341718
E-mail:newindiapublishingagency@gmail.com
www: www.nipabooks.com
For customer assistance, please contact
Phone: + 91-11-27 34 17 17 Fax: + 91-11- 27 34 16 16
E-Mail: feedbacks@nipabooks.com

ISBN: 978-93-91383-07-7

Composed and Designed by NIPA.

Dedicated To My Father
Sri Sohit Prasad Yadav, born 1925

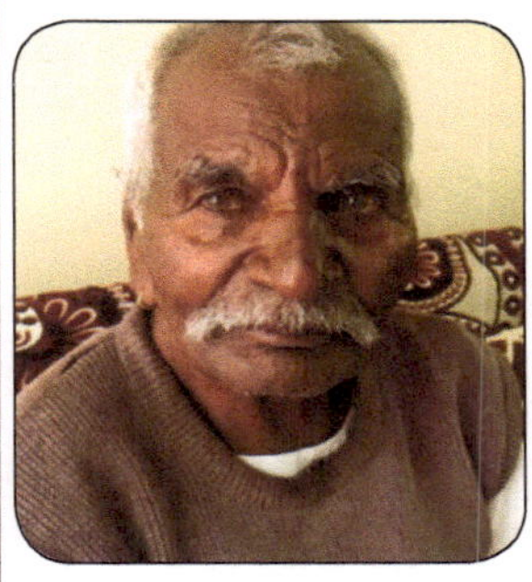

My father at the age of 96 years blessed with good health and well being taught us valuable lessons of life and continues to be a source of inspiration for us to move forward following his ideals of hard work, disciplined way of life, honesty, ethics and values.

Tribute to My Wife (Late) Smt. Chandra Yadav (1950-2019)

You departed to another world suddenly in the pleasant morning of 31 July, 2019 while taking a walk in our residential society garden. You left us stunned but then you departed the way you always wished to. You always remain besides us. We cherish your sweet memories. Your aroma of blissful presence permeates all our senses.

Hari Har Ram & Family
31 July, 2021

Preface

The present book entitled "Commercial Plant Breeding-II-Field Crops" is a follow-up book after "Commercial Plant Breeding-I-Vegetable Crops" and is meant for the course Commercial Plant Breeding listed under elective courses of State Agricultural Universities. The contents of this book include plant breeding approaches and seed production procedures with emphasis on hybrid seed production and related activities as practiced by seed companies across major field crops. The proposed course as per ICAR-V Dean Committee 'Commercial Plant Breeding' includes both field and vegetable crops in the same course. However, it was considered prudent to prepare two books, one dealing with vegetable crops and the other dealing with field crops, separately as nationally and internationally these two groups of the crops are dealt with separately at all levels. Even, private seed companies and ICAR institutes and state agricultural universities deal with these two groups of crops separately for teaching/research purposes. Therefore, it was agreed with the publisher that there will be two books; the first dealing with vegetable crops and the second dealing, with field crops and that is how 'Commercial Plant Breeding-2-Field Crops' has been prepared in continuation with 'Commercial Plant Breeding-II-Vegetable Crops' which is already out.

Initial 11 chapters under Part-I of the book are devoted to defining commercial plant breeding and outlining historical perspectives; field crops seed business, organizational structure of commercial plant breeding programme and few most important seed related acts and those related to IPRs. After these chapters, the book coverage shifts to Part-II of book having 14 chapters dealing with the breeding of individual major field crops which make business sense to the private seed companies and are of national/international importance. In crop-wise chapters on breeding of individual field crops the focus is on product design as per market requirements, commercial breeding goals, commercial plant breeding approaches highlighting hybrid breeding and hybrid seed production on commercial scale. An adequate literature review covering landmark publications from ancient times to the present has been included. A comprehensive glossary is another added feature of this book.

I am sure that this book will be accepted by the undergraduate agriculture students and also the postgraduate students reading courses on field crop breeding and will be a handy book to the research professionals both in the public and the private sectors.

31 July 2021 **(Hari Har Ram)**

Acknowledgements

Having done Masters and Ph.D. degrees in Plant Breeding from the G.B. Pant University of Agriculture and Technology (the then UP Agricultural University), Pantnagar, I started my professional career as a Junior Vegetable Breeder at Agricultural University Pantnagar in 1973 and was placed in the department of Plant Breeding. This deviation of placing vegetable breeder in the department of plant breeding rather than in the department of Horticulture happened due to foresightedness of Dr. Dhyan Pal Singh, the then Vice-Chancellor; Dr. N. K. Anant Rao, the then Dean Agriculture; Dr. R. L. Paliwal, the then Director Research and Dr. V. L. Asnani, the then Head of Department of Plant Breeding at Agricultural University, Pantnagar. This arrangement gave me a platform to practice vegetable breeding while being in the department of Plant Breeding. Later on soybean breeding was added to my responsibilities by Dr. B. P. Pandya, the then Head, Department of Plant Breeding, Pantnagar and I got huge exposure to vegetable and soybean breeding. I am thankful to all of them.

My sincere appreciation is to Dr. H.S. Gupta, former Director, IARI and former Director General, Borlaug Institute for South Asia; Dr. Kamendra (Ken) N. Mishra, Chairman, President and CEO, Pro-Farm Seed, India, Pvt. Ltd. Secunderabad; Dr. S.N. Nigam, former Principal Breeder, Groundnut Breeding, ICRISAT, Hyderabad; Dr. R. S. Mahala, Director, R&D Operations, Dupont, APAC Region, Hyderabad; Dr. R.K. Singh, ex-Rice Breeder, International Rice Research Institute, Manila, Philippines, and currently Project Leader of Crop Diversity and Genetics at International Centre for Bio-saline Agriculture at Dubai; Dr. Manju Vishwakarma, General Manager, R and D, Ankur Seeds, Nagpur; Dr. S. K. Tikoo, Co-Founder and Director, Tierra Seeds, Hyderabad; Dr. Pawan Kumar Singh, Head, Wheat Pathology, Maize and Wheat Improvement Centre, Mexico; Dr. Shiv Kumar Agrawal, Grain Legume Breeder, ICARDA; Dr. Arti Singh, Assistant Prof. Plant Breeding, Iowa State University, USA; Sri Ramashrya Yadav, former Coordinator, Seed Production Research, Limagrain, Hyderabad; Sri Amar Yadav, Managing Director, Mansoon Seeds, Pune; Dr. Kunwar Harendra Singh, Principal Scientist (Plant Breeding), ICAR-Directorate of Rapeseed Mustard Research, Bharatpur, Rajasthan; Dr. Ram Bhajan Verma, ex-Professor Plant Breeding and Chief Oilseeds Breeder, GBPUAT, Pantnagar; Mr. Pradeep Kumar, Scientific Officer, Pearl Millet Breeding, ICRISAT, Hyderabad; my elder son, Dr. Rakesh Yadava (Ph. D. Genetics and Plant Breeding), Head of India R&D Operations, Bayer Crop Science-Vegetable Crops, Bangalore; my younger son, Mr. Ranjan Yadav, Chief Manager, Bank of Baroda, Lucknow; my younger nephew, Er. Vaibhav Yadav; my younger brother; Er. K. P. Yadav, (retired) Chief Engineer, PWD, UP and currently Advisor, Uttar Pradesh State Highway Authority, Lucknow who all helped and supported in their own ways to bring this book in the present form. Special thanks to my grand-daughter Gauri for helping in photographic work.

While preparing the manuscript of this book, I had to stay a bit long at Bangalore. I am thankful to my close family members, namely, sons Rakesh and Ranjan, daughters in law Poonam and Sunita and grand children Aakash, Dhara, Gauri and Vidhi for making my stay hassle free and joyful and thus providing me an environment conducive to carry out part of this assignment smoothly and successfully.

31 July 2021 **(Hari Har Ram)**

Contents

PART 2
COMMERCIAL BREEDING OF FIELD CROPS

Glossary

A: Abbreviation for adenine (Ade. 6 - aminopurine), purine base characteristic component of nucleic acid.

AAD: See arbitrarily amplified DNA.

Allele mining: Identification and access to allelic variation of a sequenced gene that affects the plant phenotype. The DNA sequence or parts of it are used for searching new allelic variants of the same (resistance) gene.

A line: The male sterile parent line in a cross to produce hybrid seed. The A line is used as female for hybrid seed production.

Abaxial: On the side of an organ away from the axis (= dorsal).

Aberrant: Departing from the normal.

Abort: The collapse of seeds prior to maturation.

Abortive: Defective or barren.

Abscissa: The horizontal scale on a graph.

Abundance: The average number of molecules of a specified mRNA or protein in a given cell at a given time.

Accessions: Individual plant entries in world collections.

Accession number: Each accession is assigned a particular number, for example, IC (indigenous collection) numbers in National Bureau of Plant Genetic Resources gene bank and PI (plant introduction) numbers in the USDA gene bank.

Acclimatization: The adaptation of an individual to a changed climate, or the adjustment of a species or a population to a changed environment over a numbers of generations.

Acentric fragment: A piece of chromosome lacking a centromere.

A-chromosome: One of the standard chromosomes of an organism.

Acrocentric: A modifying term for a chromosome or chromatid that has its centromere near the end.

Acropetal: Toward the apex. The opposite is basipetal.

Acrydine dyes: A class of positively charged polycyclic molecules that intercalate into DNA and induce frame shift mutations.

Actinomorphic: Sepals and petals radiating from the centre of the flower; the sepals are all similar to one another, as are the petals to one another.

Active collection: collection of germplasm used for regeneration, multiplication, distribution, characterization, and evaluation, maintained in sufficient quantity to distribute on request to other scientists.

Adaptation: The process by which individuals (or parts of individuals, populations or species) change in form or functions in such a way to better survive under given environmental conditions. Also the result of this process.

Adaptive complexes: The complement of genes that adapt a plant to its environment.

Adaptor primer: A synthetic oligodeoxynucleotide that functions as a primer for reverse transcription or amplification by for example the polymerase chain reaction, and carries one or several restriction sites.

Adaptor: A synthetic oligonucleotide that contains two or more restriction enzyme cleavage sites.

Adaxial: Toward the axis (= ventral).

Addition line: An addition line has one pair of chromosome from another variety or species in addition to the normal somatic chromosome complement (2n) of the species.

Additive gene effects: Gene action in which the effects on a genetic trait are enhanced by each additional gene, either an allele at the same locus or genes at different loci.

Additive genes: Gene interaction without dominance (if allele) or without epistasis (if non-allele).

Adelphogamy: Sib pollination.

Adenine: A purine base found in RNA and DNA.

A-DNA: A right handed DNA double helix that has 11 base pairs per turn. DNA exists in this form when partially dehydrated

ADP: Initials for adenosine diphosphate, a complex sugar-phosphate compound formed as a result of energy expenditure and loss of a phosphate group from the energy rich ATP (adenosine triphosphate) compounds.

Adult plant resistance: Resistance only visible in the adult stage of a plant, i.e., at the generative phase (contrast seedling resistance). Adult-plant resistance can be inherited mono- or oligogenically. Adult-plant resistance need not be durable.

Advance breeding line: Plants whose germplasm has been manipulated to form the parental material from which new varieties or hybrids are evolved.

Adventitious: Describing buds, roots, etc. which develop in an irregular or unusual position.

AFLP (amplification fragment length polymorphism): A variant of DNA amplification product of different size generally produced by amplification with arbitrary primers (e.g. in AFLP, AP-PCR, DAF, RAPD). The term was originally coined to depict DAF polymorphism. AFLP can be used to discriminate between related individuals, localize specific gene or gene regions in complex genomes and establish genetic maps. An arbitrarily amplified DNA technique that uses ligation of adaptors to the ends of restricted DNA and amplification with hemi-specific primers. Amplification products are generally visualized by polyacrylamide gel electrophoresis and autography, and fingerprints are highly complex. The AFLP method is also known by selective restriction fragment amplification (SRFA).

Agamic: Without sex, asexual.

Agar: A dried sea-weed extract from red algae such as Geledium and Gracilaria, consisting of a mixture of agarose and agaropectin. Agar or agar-agar is used to solidify media in bacteriology and tissue culture.

Agarose: A neutral, linear sulphated galactan polysaccharide used extensively in electrophoresis.

Agarose gel: A polysaccharide matrix for electrophoretic separation of nucleic acid molecules according to size and conformation. These gels are inert, macroporous, and nontoxic. Agarose gels are easy to cast but lack the resolution of polyacrylamide gels and cannot be easily preserved. DNA is usually stained in agarose gels with dye ethidium bromide.

Agriculture: The science of transforming sunlight energy into products that can be stored and used by humans at a later date. Plants that photosynthesize are the primary converters and the animals that consume the products of photosynthesis, the secondary.

Agrobacterium: Soil living bacteria genus which can transfer plasmids into plants and other organisms.

Agrobiodiversity: The variety and variability of animals, plants, and microorganisms used directly or indirectly for food and agriculture; it comprises the diversity of genetic resources and species used for food, fuel, fodder, fibre and pharmaceuticals.

Agrobiology: The study of plant life in relation to agriculture, especially with regard to plant genetics, cultivation and crop yield.

Agrobiotechnology: Modern biological knowledge and methods that can be applied to human goals in agriculture.

Agroecosystem: A comprehensive approach to factors associated with agriculture and ecology and their interactions.

AIDS (acquired immunodeficiency syndrome): The usually fatal human disease in which the immune system is destroyed by the human immunodeficiency virus (HIV).

Albino: Mutant form of plants showing degrees of chlorophyll deficiency ranging from complete to banding.

Albuminous seed: A mature seed which contains endosperm.

Aleurone: The outermost layer of the endosperm in a seed.

Alien germplasm: Genes introduced from a wild relative or un-adapted species.

Allele: An allele is an alternative form of a gene.

Allele frequency: The proportion of one allele relative to all alleles at a locus in a population.

Allele mining: An approach to access new and useful genetic variation in crop plant collections.

Allelopathy: Release of growth-inhibiting substances from higher plants.

Allogamy: When pollen grains from flowers of one plant pollinate the flowers of other plants (cross pollination).

Allogene: Recessive allele.

Allopatric speciation: Speciation occurring at least in part because of geographic isolation.

Alloploid (allopolyploid): An organism with more than two sets of chromosomes in the somatic cells. The additional sets having been derived from different species.

Allotetraploid: An organism with four genomes derived from hybridization of different species. Usually in forms that become established, two of the four genomes are from one species and two are from another species.

Allozygote: A diploid individual in which two genes at a particular locus are not identical by descent from a common ancestor.

Allozyme: Enzymes produced by genes at several alleles.

Alternate (of leaves, etc.): Not opposite to each other on the axis, but borne at regular intervals at different levels.

Amino acid: Any one of a class of organic compounds having an amino (NH_2) group and a carboxyl (COOH) group. Amino acids are the building blocks of proteins.

Amorph: A mutation that obliterates gene function (a null mutation).

Amphiploid (amphidiploid): An alloploid possessing the total chromosome complements of the parental species.

Amplicon: Nucleic acid region defined by two oligonucleotide primer annealing sites in opposite strands.

Amplification: An increase in copy number of a particular DNA fragment, resulting from replication of a vector in which it has been cloned, replication of specific gene in absence of general chromosomal replication, or in vitro enzymatic activity generally driven by short oligonucleotide segments (primers).

Amplimer (amplification primer): A synonym for oligonucleotide primer.

Analysis of variance: Partitioning variation among data into its components by the computation of suitable statistics.

Anaphase: A stage of mitosis or meiosis during which the daughter chromosomes pass from the equatorial plate to opposite poles of the cell (towards the ends of the spindle). Anaphase follows metaphase and precedes telophase.

Anchor gene: A gene that has been positioned on both the physical map and the genetic map of the chromosome.

Anchored primer: A specifically designed primer which contains a region with high sequence homology to specific nucleic acid region, such as a repetitive motif, poly (dA) tail, etc. Primers can be anchored at their 5` and 3` termini.

Androdioecy: Mating system where some individuals have male flowers and some have hermaphrodite flowers.

Andromonoecy: Mating system where every plant possesses both male and hermaphrodite flowers.

Aneuploid: An individual with other than an exact multiple of the haploid chromosome complement.

Angiosperm: Flowering plants whose seeds are enclosed within an ovary.

Anisoploid: Mixture of diploid, triploid, and tetraploid plants obtained with seed harvested from a mixture of diploid and tetraploid plants.

Annealing: A spontaneous alignment of two single stranded nucleic acids to form a duplex molecule. The alignment interaction is based on hydrogen bond formation between complementary base pairs in the individual strands and the formation of a double helix.

Annuals: Plants that complete their life cycle from seed in one year.

Anther: The polliniferous part of the stamen.

Anther culture: The culturing of anthers in vitro for the purpose of generating haploid plantlets.

Anthesis: Full flower expansion including anther extrusion.

Antibiosis: Antagonistic association in which one organism has an injurious effect on normal growth and development of another.

Anticodon: A three nucleic acid sequence or transfer RNA that complements a messenger RNA codon in the formation of protein.

Antigen: A substance usually a protein that is bound by an antibody or a T cell receptor when introduced into a vertebrate organism.

Antisense gene: A gene that produces a transcript that is complementary to pre-mRNA or mRNA of a normal gene (usually constructed by inverting the coding region relative to promoter).

Antisense RNA: RNA that is complementary to the pre-mRNA or mRNA produced from a gene.

Apical: Relating to the apex or tip.

Apical dominance: In plants, the inhibiting of lateral bunds by high levels of auxins produced in the lead shoot or apical meristem.

Apogamy: Development of embryo from synergids or antipodal cells without fertilization; a form of apomixes.

Apomixis: Reproduction from an unfertilized egg or from somatic cells associated with the egg.

Apospory: A form of apomixes in which the embryo sac develops from a vegetative cell of the ovule.

Appressed: Lying close and flat against erect type.

Approach crossing technique: The pollen donor inflorescence is put slightly above the emasculated inflorescence and bagged or enclosed in opposite ends of a plastic sleeve.

Arabidopsis thaliana: A small cruciferous weed of mustard family with a small genome, well characterized genetics and advantageous cultural characteristics. The plant has been used as a model organism in molecular biological research.

Arbitrarily amplified DNA (AAD): Collective term to describe those nucleic acid amplification based techniques that use arbitrary primers. Also termed AFLP analysis. Arbitrary primer technology (APT), multiple arbitrary amplicon profiling (MAAP), and nucleic acid scanning by amplification.

Arbitrarily primed PCR (AP-PCR): An arbitrarily amplified DNA technique that uses one or more primers of typically between 18 and 32 nucleotides in length and high primer-to-template ratio (1-500). Amplification products are generally visualized by polyacrylamide gel electrophoresis and autoradiography, and the fingerprints are moderately complex.

Arbitrary sequence oligonucleotide fingerprinting (ASOF): A non-targeted hybridization strategy collection of sequence tagged fragments.

Articulate: Jointed: having a node or joint.

Artificial seed: A single somatic embryo in a protective coating that mimics the size, shape and physiological condition of true seed.

Artificial selection: Elimination of segregating plants by mechanical, chemical, or mechanical human selection.

ASAP (allele specific associate primer): A synthetic oligodeoxynucleotide of 20-30 nucleotides, that is complementary to the 3` or 5` end of an arbitrarily amplified DNA product and is used in a PCR reaction to amplify a specific allele.

Asexual reproduction: Any process of reproduction that does not involve the formation and union of gametes from different sexes or mating types.

Asynapsis: Failure of pairing of homologous chromosomes during meiosis.

Auricle: An ear-shaped appendage or lobe.

Auriculate: Furnished with auricles.

Autogamy: Self fertilization.

Autoploid (autopolyploid): An organism with more than two sets of chromosomes in the somatic cells, the sets derived from the same species.

Autoradiography: Method or documentation used to detect radioactively labeled molecules superimposed to a photosensitive emulsion or film.

Avirulent: Parasite unable to infect and cause disease in a host plant.

Axil: The angle formed between any two organs.

Axillary: In or related to the axis.

Axis (of stem, inflorescence, etc.): The central part of a longitudinal support in which organs or parts are arranged.

B line: The fertile counterpart, or maintainer of the "A" line. The B-line does not have fertility restorer genes and is used as the pollen parent to maintain "A" line.

Backcross: A cross of a hybrid with one of its parents or with a genetically equivalent organism.

Back mutation: A second mutation at the same site in a gene as the original mutation which restores the wild type nucleotide sequence.

Bacteria: Bacteria are single cell, microscopically small organism with simple structure usually reproduce asexually with very short generation time (20 min.). A single bacterium produced clones are genetically identical. Accidental changes (mutation) lead to new type of clone.

Bacterial artificial chromosome (BAC): A bacterial cloning vector based on a single copy F-factor of *Escherichia coli* that allows the cloning of fragments of an average size of up-to 300 kb.

Balance: The condition in which genetic components are adjusted in proportion that give satisfactory development. Balance applies to individuals and populations.

Balanced polymorphism: Two or more types of individuals maintained in the same breeding population.

Base: A heterocyclic nitrogen containing molecule constituent of DNA and RNA. The most common purine and pyrimidine bases are adenine, guanine, cytosine, thymine and uracil.

Base collections: In gene banks cold storage, this is the primary or long term seed source, in which germination is renewed after appropriate intervals.

Base composition: The ratio of number of adenine and thymine (AT) to guanine and cytosine (GC) bases in a DNA molecule. In double stranded DNA, A= T and G=C.

Basic number of chromosome: The number of chromosomes in ancestral diploid ancestors of polyploidy, represented by x.

BC_1, BC_2, etc: Symbols used to designate the first backcross generation, the second backcross generation, etc.

Biennial: A plant that requires two years to complete its life cycle from seed to seed.

Biochemistry: Chemistry of living organisms.

Biodiversity: Number and relative composition of species or ecotypes in a certain habitat. The existence of a wide amount of diversity present in nature.

Biological species concept: A system in which organisms are classified in the same species if they are potentially capable of interbreeding and producing fertile offspring.

Biological value: An index of the portion of absorbed nitrogen retained in the body for growth or maintenance or both.

Biological yield: The total yield of plant material.

Biomass: Renewable organic matter formed through the process of photosynthesis.

Biometry: The branch of science which deals with statistical procedures in biology.

Biotechnology: The application of recombinant DNA, cell and tissue culture, and other methods used to develop new and improved plants and plant products.

Biotype: (1) A population in which all individuals have an identical genotype; (2) a physiological race.

Bivalent: A pair of homologous chromosomes united in the first meiotic division.

Blade joint: The flexible union between the leaf blade and the leaf sheath.

Blade: The expanded portion of a leaf, etc.

Bloom: A whitish, powdery and glaucous covering of the surface, often of a waxy nature.

Blot: A nitrocellulose or nylon membrane to which a nucleic acid or protein has been transferred or the autograph of a blotting experiment.

Blunt end: A non-cohesive end which is perfectly base-paired terminus of a duplex DNA fragment.

Bolting: Production of seed stalks.

Botany: The science of biology that deals with plant life.

Bract: A more or less modified leaf subtending a flower or belonging to an inflorescence.

Bran: The outer layers of a cereal grain removed in milling, consisting of the pericarp, testa and the aleurone layer.

Brand: A commercial name applied to seed.

Breeder seed: Seed increased by the originating, or sponsoring, plant breeder or institution, used as the source for production of foundation seed.

Breeders' rights: Legislation giving the equivalent of patent rights for plant cultivars. The effect is to give the developer of a cultivar sole legal possession of that cultivar and a legal basis for compensation for its use by others. Breeder`s Rights may be referred to as `Plant cultivar protection'.

Breeding: The art and science of improving plants or animals genetically.

Breeding value: In quantitative genetics, the part of the deviation of an individual phenotype from the population mean, that is due to the additive effects of alleles.

Bridge cross: A method of overcoming the incompatibility between two species, the F_1 of one cross is used as bridge to hybridize third species.

Bridging species: A species used in gene transfer from one species to another sexually incompatible species; bridging species is compatible with both the donor and recipient species, it may be a natural or synthetic species.

Broad sense heritability: In quantitative genetics, the proportion of the total phenotypic variance that is the genotypic variance.

Broad spectrum resistance: Individual locus/QTL that confers resistance to multiple races of a pathogen species or multiple taxa of pathogens.

BSA: Abbreviation for bovine serum albumin/bulk segregant analysis.

Bt. (Bacillus thuringiensis): A naturally occurring bacterium that produces a protein that is toxic to certain lepidopteran insects.

Bud: The rudimentary state of a stem or branch.

Bulb: An underground storage organ, containing a short, flattened stem with roots on its lower surface, and above its fleshy leaf base, surrounded by protective scale leaves.

Bulk method: In this method F_2 and the subsequent generations are grown in bulk, usually without artificial selection.

Bulk segregant analysis: Detection strategy used to define molecular regions controlling a certain phenotype. Large populations of organisms are usually pooled (bulked) into two classes representing distinguishing phenotypes. DNA profiling with arbitrary primers (RAPD, DAF) is then used to generate DNA markers associated with those regions.

Bulliform: Applied to large, thin walled epidermal cells of most Gramineae.

Buzz-pollination: Type of pollination where bees vibrate flowers for collecting pollen.

C: Abbreviation for cytosine (6-amino-2-hydroxy-pyrimidine), a pyridine base characteristic of DNA and RNA.

CAAT box: A conserved nucleotide sequence in eukaryotic promoters involved in the initiation of transcription.

Callus: A disorganized mass of undifferentiated plant cells obtained through culture from small pieces of plant tissues or a mass of undifferentiated cells, originating from an anther microspore, plant tissue explant, or other cellular source, when cultured in vitro.

c-DNA (complementary DNA): A single or double stranded DNA molecule complementary to an RNA, generally mRNA made by RNA-dependent DNA polymerase (reverse transcriptase).

c-DNA library: A collection of c-DNA clones containing copies of the RNAs isolated from an organism or a specific tissue or cell type of an organism.

C_4 plants: plants having c_4 (dicarboxylic acid) photosynthetic pathway for example corn, sorghum, millets.

CAPS: Cleaved amplified polymorphic sequence.

Carpel: A simple pistil, or one member of compound pistil.

Caryopsis: A grain, as in grasses; a seed-like fruit with a thin pericarp adnate to the contained seed.

Cell: The smallest membrane bound living unit of biological structure capable of auto-reduplication.

Cell culture: A process of culturing and growing cells through in vitro and regenerating plants by several sub-cultures.

Centi-Morgan (cM): Unit of recombination named after the geneticist Morgan. One cm is equivalent to one per cent recombination and is a measure of relative distance between two genes or genetic markers in a chromosome.

Centre of diversity: Areas where cultivated plant species and/or their wild relatives show much greater variation than anywhere in the rest of the world.

Centre of domestication: The area believed to be that in which a particular crop species was first cultivated.

Centromere: A localized region in each chromosome to which spindle fibres appear to be attached.

Cereal grain: Grasses cultivated primarily for their edible seeds or grains.

Certified seed: The progeny of foundation or certified seed, produced and so handled as to closely maintain genetic purity or identity of a variety. Certified seed is approved and certified by an official seed certification agency.

Character: The expression of a gene as revealed in the phenotype. This is phenotypic trait of an organism controlled by genes and may be dominant or recessive and may be governed by single gene or multigenes.

Chasmogamy: Fertilization after opening of flowers.

Check cultivar: A standard reference cultivar for test comparisons.

Chemical hybridizing agents: Compound applied to plants prior to anthesis to selectively induce male sterility.

Chiasma: Crossing over resulting in an interchange at corresponding points between homologous chromosomes.

Chimera: An individual that is composed of cells of two or more genotypes.

Chimeric selectable marker gene: A gene that is constructed from parts of two or more different genes and allows that host cell to survive under conditions where it otherwise would die.

Chi-square: A statistic used to test the hypothesis of fit to the predictions of a hypothesis.

Chlorophyll: The major photosynthetic green pigment present in the chloroplast.

Chromatid: One of two thread like structures formed in the duplication of a chromosome to form daughter chromosomes.

Chromatography: A process of separation of materials into bands of different colours based on preferential adsorption of constituents on strip of filter paper or column of silica.

Chromosome: A structural unit in the nucleus, which carries the genes in a linear constant order, it preserves its individuality from one cell generation to the next and is typically constant in number in any species.

Chromosomal (genome) walking: A technique for the sequential isolation of clones present in a genomic library with sequences that overlap and span large chromosomal intervals.

Chromosome aberration: The lack of complete chromosome complement.

Chromosome jumping: A procedure that uses large DNA fragment to "jump" along a chromosome from one site to another site.

Chromosome substitution: The process of replacing one pair of chromosomes of a variety with those of another variety of a related species.

Cistron: A functional unit of gene which might be broken by recombination.

Clavate: A term to describe a spike that is compacted and enlarged at the top, giving a club-shape appearance.

Cleaved amplified polymorphic sequence (CAPS): A technique for the detection of restriction fragment length polymorphisms at a specific genome locus originally amplified by the polymerase chain reaction. The technique uses restriction endonucleases to cut the amplified DNA and identify sequence polymorphism.

Cleistogamy: Pollination and fertilization in an unopened flower bud.

Cline: Graded series of a morphological or genetic traits, distributed along a spatial dimension.

Clonal selection: A method of selection based on clones. Does not involve sexual reproduction.

Clone: Individuals obtained from a single plant through asexual reproduction (produced by mitotic cell division).

Cloning vector: Any nucleic acid molecule capable of autonomous replication in a host cell and capable of accepting foreign DNA.

Cloning: The use of recombinant DNA technology to propagate the sequence of a DNA fragment in a cloning vector, after its transformation into a suitable host cell.

Cluster analysis: The analysis and characterization of subsets of organisms (operational taxonomic units) through search of properties of coherence and isolation.

Co-dominance: The phenomenon whereby both the alleles at a locus are expressed and influence the phenotype of the heterozygote.

Cob: The central axis of an ear of corn on which kernels develop.

Codon: A particular triplet sequence of bases in DNA specifying an amino acid.

Coefficient of variation: The standard deviation expressed as a percentage of the mean.

Cohesive ends: The termini of two DNA duplex molecules capable of annealing to each other.

Colcemid: A synthetic equivalent of colchicine.

Colchicine: An alkaloid extracted from seeds or corms of *Colchicum autumnale* with induces polyploidy by arresting spindle formation during mitosis.

Coleoptile: A membranous cover over the shoot apex of cereal crops that protects the plumule during germination.

Combining ability, general: The average or overall performance of a genetic strain in a series of crosses.

Combining ability, specific: The performance of specific combinations of genetic strains in crosses in relation to the average performance of all combinations.

Commercial plant breeding: Commercial plant breeding refers to plant breeding (genetic improvement of crop plants suiting to the requirements of the human beings) with focus on products as per requirements of the market. This is commonly practiced by the commercial seed companies using more input of science and data and less and less of art component in taking decisions in time-bound frame so that the commercial products reach to the market in the least possible time following a rapid breeding cycle in a cost effective manner.

Compensation point: The level of photosynthetically active radiation or CO_2 required to equal respiration losses.

Competence: Ability of a bacterial cell to incorporate DNA and to become genetically transformed.

Complexity: The length of non-repetitive base sequences in a nucleic acid molecule.

Complete flower: A flower has sepals, petals, stamens and pistils.

Composite: A mixture of genotypes from several sources, maintained by open pollination.

Conditioning: The term used to describe the process of cleaning seed and preparing it for market, previously called as seed processing.

Conoidal: Nearly conical.

Consensus (conserved) sequences: A defined nucleotide sequence either frequent at a defined position or characteristic for a specific functional region of a gene, occurring generally in the same context in a set of DNA sequences.

Constitutive gene: A gene that is continuously expressed in all cells of an organism.

Contig: Contiguous segment of DNA used in genome analysis and defined as a group of genomic clones that contain mutually overlapping sequences.

Controlling element: In maize, a transposable element such as Ac or Ds that is capable of influencing the expression of a nearby gene.

Convergent improvement: Backcrossing of a single cross independently to both of its inbred parents with the view of improving both the inbreds.

Core collection: The basic sample of a germplasm collection; designated to represent the wide range of diversity in terms of morphology, geography range, or genes.

Correlation: A mutual relationship between two things such that an increase or decrease of one is generally associated with an increase or decrease of the other. Linear correlation is measured by the correlation coefficient, which may range in value from - 1 to 1.

Cortex: Rind or bark.

Cosmids: Cloning vectors that are hybrids between phage lambda chromosomes and plasmids, they contain lambda cos sites and plasmids origin of replication.

Cotyledon: The foliar portion of first leaves (one, two, or more) of the embryo as found in the seed.

Coupling: Linked recessive alleles occur in one homologous chromosome and their dominant alternatives occur in the other chromosomes. Opposed to repulsion in which one dominant and one recessive occur in each member of the pair of homologous chromosomes.

Critical leaf area: A leaf area above which detrimental mutual shading reduces net photosynthesis rate.

Crop ecology: The study of the relationship between crop plants and their environment.

Cross breeding: Out-breeding.

Cross fertilization: Fertilization by pollen from another plant.

Cross over value: The percentage of crossing over in a hybrid population; a term used mostly in determining linkage percentage, particularly in chromosome mapping.

Cross protection: Is a phenomenon where infection of a plant with one viral strain protects against supper infection by another related strain.

Cross resistance: Resistance associated with a change in one genetic factor that results in resistance to different chemical pesticides that were never applied.

Crossing over: An interchange of segments between the non-sister chromatids of two homologous chromosomes at meiosis.

Cryopreservation: Seed storage at -196°C, the temperature of liquid nitrogen.

Cultigen: A cultivated genotype, as contrasted to wild species.

Culm: A commonly used name for the stem of grass plant, consisting of nodes and usually hollow internodes.

Cultivar: A contraction of the words cultivated varieties, equivalent to variety.

Cut: A dnase induced break in a double-stranded nucleic acid molecule, as opposed to nick, a beak in a single stranded molecule.

Cuticle: The outer film of epidermal cells, impermeable to water.

Cybrid: Hybrid developed with fusion of the cytoplasm of two cells/parents.

C-value: The total amount of DNA in a haploid genome.

Cytogenetics: A hybrid science that deals with the study of cytology (study of cells) and genetics (study of inheritance). It includes the study of number, structure, function and movement of chromosomes and numerous aberrations of these properties as they relate to recombination, transmission, and expression (phenotype) of the genes.

Cytokinins: A group of plant hormones that promote primarily cell division as well as cell enlargement, differentiation, dormancy, and retardation of leaf senescence.

Cytology: A study of cell structure and components in relation to their function.

Cytoplasm: The protoplasm of a cell excluding the nucleus.

Cytoplasmic: Pertaining to or centered in the cytoplasm.

Cytoplasmic-genetic male sterility: Cytoplasmic male sterility for which a restorer gene is available.

Cytoplasmic inheritance: Inheritance dependent upon hereditary units in the cytoplasm. It is inheritance of traits whose determinants are located in the cytoplasmic organelles, also known as non-Mendelian inheritance.

Cytoplasmic male sterility: Sterility due to factors that are carried only through the female and are not diluted or lost in successive generations of reproduction.

Cytosine: A pyrimidine base found in RNA and DNA.

DAF: DNA amplification fingerprinting.

Data analytics: Coupled with broader technological improvements into data gathering and analysis, the process by which genes are selected and new crops make it into fields and onto consumers table is more efficient than ever before. If one can use data to make a better decision today about which corn hybrids to produce over the winter then that can lead breeders to launch a new commercial product much faster.

Day neutral plants: Plants that have no day-length requirements for floral initiation.

Deep water or floating rice: Rice plants that can elongate rapidly when flood water rises and that have adventitious roots at submerged nodes.

Deficiency: The absence or deletion of a segment of chromosome.

Dehiscence: Splitting open of a fruiting structure or anther.

Deletion: The loss of a chromosome segment involving one or more genes.

Deltoid: Shaped like the Greek letter delta.

Denaturation: The loss of native configuration of a macromolecule. In particular, DNA denaturation refers to the conversion of double stranded to single stranded molecules by melting, breaking hydrogen bonds.

Dendogram: An evolutionary tree diagramme may order objects, individual genes on the basis of similarity.

Denovo: Latin word meaning arising a new, a fresh.

Detasselling: Removal of the tassel (male inflorescence) as in maize.

Determinate: Of plants, exhibiting clearly defined vegetative growth and reproductive development stages where terminal bud terminates in an inflorescence.

Deviation: Departure of an observation from its expected value.

Diallel cross: The crossing in all possible combinations of a series of genotypes.

Diagnostic marker: Markers that could be used in multiple genetic backgrounds, ideally the marker-trait association is valid in all germplasm.

Dichogamy: Maturation of male and female reproductive organs of a hermaphrodite flower at different times.

Dicotyledonous: Having two cotyledons or two sections of a seed that form two cotyledonary leaves at the first node.

Differential cultivar: A cultivar that contains known genes for disease reaction used for identification of physiological races of pathogens.

Dihybrid: Heterozygous with respect to two genes.

Dioecious: Plants in which staminate and pistillate flowers occur on separate plants.

Dioecy: State of being dioecious.

Diploid: An organism having two genomes i.e. with chromosome number of 2x.

Diplospory: A kind of apomixes in which the embryo sac develops from the megaspore.

Diplotene: The stage of meiosis which follows pachytene and during which the four chromatids of each bivalent move apart in two pairs but remain attached in the region of chiasmata.

Disease: Any disturbance in a plant that interferes with its normal structure and function or its economic value.

Disjunction: The separation of chromosomes at anaphase.

Disruptive selection: Selection of extreme classes of plants and inter-mating them to maintain polymorphism.

Disomic: An individual with two homologues for each chromosome of the genome.

Distant hybridization: The crossing of species of different genera.

DNA: Deoxyribonucleic acid, the chemical basis of heredity. DNA (deoxyribonucleic acid), a single or double stranded polymer of nucleotides arranged along a deoxyribose and phosphate backbone capable of carrying genetic information in most organisms.

DNA amplication fingerprinting (DAF): An arbitrarily amplified DNA technique that uses one or more primers of typical between 5 and 15 nucleotides in length and high primer-to-template ratios (5-50,000). Primers can harbour secondary structures, such as hairpins or mini-hairpins. Amplification products are generally visualized by polyacrylamide gel electrophoresis and silver staining, and fingerprints are highly complex.

DNA array (chips): A tool for the analysis of nucleic acid sequence that uses arrays of immobilized nucleic acid molecules in hybridization experiments, sometimes coupled with amplification. A number of variants have been described whereby the target or the probe molecules (oligonucleotide) are bound to a solid support.

DNA fingerprinting: The method that generates a DNA fingerprint or pattern capable of detecting DNA polymorphism. Typically, genomic DNA is cut with restriction endonuclease, and the resulting DNA fragments are separated by gel electrophoresis, transferred to a membrane, and hybridized with a labeled fingerprint probe consisting of either a synthetic oligonucleotide or a cloned DNA fragment. In microsatellite fingerprinting, the DNA fragments can be directly hybridized in the gel thus doing away with the membrane transfer.

DNA polymorphism: The variation in length, presence or sequence of a DNA fragment generally produced by restriction or amplification. These polymorphisms are generated by mutation, insertion, deletion or rearrangement of DNA, and are particularly evident in hyper-variable genomic regions, such as microsatellite and mini-satellite sequences.

DNA sequencing: Any of several methods capable of determining the sequence of bases in a DNA molecule.

DNA synthesis: The chemical or enzymatic linking of deoxyribonucleotide triphophates to a nascent oligonucleotide polymer.

DNase: An enzyme that catalyses the hydrolysis of phosphodiester bonds in single or double-stranded DNA, beginning from either the termini (exonucleases) or from within (endonuclaeses) the nucleic acid molecule.

dNTP: Abbreviation for any 2-deoxynucleoside-5-triphosphate.

Domestication: The process of bringing a wild species under human management. The selective breeding by humans of species in order to accommodate human needs.

Dominance: Intra-allelic interaction such that one allele manifests itself more or less, when heterozygous, than its alternative allele.

Dominant gene effect: Gene action with deviations from the additive such that the heterozygote is more like one parent than the other.

Donor parent. The parent from which one or few genes are transferred to the recurrent parent in a backcross breeding.

Dormancy: A condition when healthy, morphologically ripe seed does not germinate or has a delayed response when placed under optimum conditions of moisture, temperature and light.

Dorsal (in the morphological sense): Upon or relating to the back or outer surface of an organ (abaxial).

Double cross: A cross between two single crosses.

Double fertilization: A phenomenon in angiosperms whereby one male nucleus unites with the egg nucleus to form a zygote (2n) which develops into the embryo and the second male nucleus unites with two polar nuclei in the embryo sac to form the endosperm (3n).

Drift: Changes in gene and genotypic frequencies in small populations due to random processes.

Drought hardiness: The capacity of plants to survive desiccation.

Drought resistance: In plants, overall suitability for cultivation under dry conditions.

Duplex: A polyploid which is recessive at all the loci except two for a particular gene (AAaa).

Duplicate genes: Two or more pairs of genes that produce identical effects, whether alone or together.

Duplication: Occurrence of a segment of a chromosome twice in the haploid set.

Durable resistance: Resistance that remains effective for a long period when applied on a large scale in a region that is undergoing regular epidemics of the pathogen.

Ear: The seed-bearing organ of the corn plant.

Ecology: The study of reciprocal relations among plants and animals and their environments.

Egg: The female gamete or germ cell.

Electrophoresis: The movement of colloidal particles through a fluid under the action of an electric field.

Electroporation: The treatment of protoplasts with high voltage electric pulse during direct gene transfer.

Elite line: An agronomically superior and high performing local cultivar.

Emasculation: Removal of the anthers from a flower.

Embryo: The rudimentary plant in a seed. The embryo arises from the zygote.

Embryo culture: The culturing of an immature embryo on a sterile nutrient medium.

Embryo sac: Typically, an eight nucleate female gametophyte. The embryo sac arises from the megaspore by successive mitotic division.

Endosperm: Triploid tissue that arises from the typical fusion of sperm nuclei with the polar nuclei of the embryo sac. In seed of certain species, the endosperm persists as a storage tissue and is used in the growth of the embryo and by the seedling during germination.

Environment: The sum of the external conditions which affect growth and development of an organism.

Enzyme: Substance produced from a gene that controls or regulates a cell function/biochemical process.

Epidermis: The superficial layer of cells.

Epigeal: Seed germination in which the cotyledons are raised above the ground by elongation of hypocotyl (beans).

Epigenetic: A term referring to the non-genetic causes of a phenotype.

Epiphytotic: An un-arrested spread of a plant disease.

Episome: A genetic element that may be present or absent in different cells and that may be conserved in a chromosome or independent in cytoplasm. For example, the fertility factor (F) in *Escherichia coli.*

Epistasis: Interaction between non-allelic genes, in which manifestation at any locus is affected by genetic phase at any other loci.

Error variance: Variance arising from unrecognized or uncontrolled factors in an experiment with which the variance of recognized factors is compared in tests of significance.

Ergot: Dark, spur shaped sclerotium that develops in place of a healthy seed in a diseased inflorescence. It is a disease of cereals and grasses. These are toxic to both human and livestock.

Escape: Phenomenon of susceptible plants avoiding disease attack.

Essentially derived variety: The concept of essentially derived variety (EDV) embodied in Article 14(5) of the 1991 Act of the UPOV Convention is designed to ensure the Convention continues to provide an adequate incentive for the plant breeding. Under that Article, a cultivar that is essentially derived from a protected cultivar may be the subject of protection (if it fulfills the normal protection criteria of DUS and novelty) but cannot be exploited without the authorization of the breeder of the protected cultivar. For practical purposes, cultivar will only be essentially derived when they are developed in such a way that they retain virtually the whole genetic structure of the earlier variety.

Ethidium bromide: A phenantridium fluorescent dye (3, 8 diamino-6ethyl-5phenyl phenantridium bromide) which intercalates between base pairs in double stranded nucleic acids, alters buoyancy, and fluoresces at UV wavelength of more than 300 nm.

Ethnobotany: The science that deals with the folklore knowledge of plants.

Eucarpia: Acronym for the European Association for Research on Plant Breeding.

Euchromatin: The portion of genomic DNA that remains relatively unstained and is transcriptionally active.

Eugenics: The application of the principle of genetics to the improvement of humankind.

Eukaryote: A member of the large group of organisms that have nuclei enclosed by a membrane within their cell.

Euploidy: Variations in chromosome numbers that are multiples of complete sets basic to a species.

Evapotranpiration: The combined loss of water from an area through evaporation and transpiration.

Evolution: The cumulative change in the characteristics of population of organisms related by descent occurring during the course of successive generations. There are changes in the genetic composition of a population with the passage of each generation.

Exogamy: Outbreeding.

Exons: The segment of a eukaryotic gene that corresponds to the sequences in the finally processed RNA transcript of that gene.

Exonuclease: An enzyme that digests DNA or RNA, beginning at the ends of the strands.

Exotic: A variety of species introduced from a foreign country.

Explant: A portion of tissue removed from a plant for tissue culture.

Exponential rate: Used to indicate variation in which one variable factor depends upon another resulting in an ever increasing rate.

Expressed sequence tag (EST): A synthetic oligonucleotide complementary to part of a specific mRNA and usually derived from a cDNA library or RNA fingerprints.

Expressivity: The degree of manifestation of a genetic character.

Extrachromosomal: Structures that are not part of the chromosomes, DNA units in the cytoplasm that control cytoplasmic inheritance.

Extrachromosomal inheritance: Inheritance that is not controlled by nuclear genes but by cytoplasmic organelles.

F-statistics: Short for fixation statistics, coined by Wright as a way of describing the distribution of genetic variation within and between populations in a species.

F_1: The first generation of a cross. It indicates the first finial generation.

F_2: The second filial generation obtained by self-fertilization or crossing inter se of F_1 individuals.

F_3: Progeny obtained by self-fertilizing F_2 individuals.

Factor: Same as gene.

Family: A group of individuals directly related by descent from a common ancestor.

Farmer's privilege: The most prominent issue in the *sui generis* systems involves the so called 'farmer's privilege' and 'breeder's exemption'. The traditional rights of farmers to save seed from their harvest to plant the following season is an important aspect of the *sui generis* systems and is one of the most contentious aspects of IPR in plant breeding. This practice is often described as farmer's right.

Fertility: Ability to produce viable offspring.

Fertilization: Union of an egg and a sperm (gametes) to form a zygote.

Fibro-vascular: Composed of woody fibres and ducts.

Filament: The part of a stamen which supports the anther; any thread like body.

Fingerprinting: The whole repertoire of techniques capable of generating a fingerprint.

Fitness: Fitness of a genotype is commonly measured by the number of successful offspring compared to other genotypes.

Floret: The stamens, pistil and lodicules enclosed by lemma and palea.

Flower: Reproductive structure in angiosperms composed of calyx, corolla, androecium and gynoecium.

Foliar: Relating to a leaf.

Forages: Plants grown primarily for feed in which all or nearly all of the top growth is harvested.

Forced cloning: The insertion of a foreign DNA into a cloning vector in a predetermined orientation.

Foundation seed: Seed stock produced from breeder seed by or under the direct control of an agricultural experiment station. Foundation seed is the source of certified seed.

Freeze preservation: Conditioning and preservation of plant cells, tissues, or organs at extremely low temperature, usually in liquid nitrogen.

Fruit: the seed-bearing product of a plant.

F-statistics: Short for fixation statistics coined by Wright as a way of describing the distribution of genetic variation within and between populations in a species.

Full-sibs: Individuals of a progeny with common parents.

Functional genomics: The field of study that attempts to determine the functions of all genes (and gene products) largely based on knowing the entire DNA sequence of an organism. The use of whole genome information and high-throughput tools has opened up a new field of research called functional genomics. Among its sub-disciplines, transcriptomics (the complete set of transcripts produced in a cell), proteiomics (the complete set of proteins produced in a cell) and metabolomics (the complete set of metabolites in a cell) have been used by the plant science community. The functional genomics refers to the development and application of global (genome-wide or system-wide) experimental approaches to assess gene function by making use of the information and regents provided by structural genomics. It employs huge amount of statistical and computational analysis (bio-informatics).

Fungi: Microscopic plants consisting of a vegetative structure called a mycelium, lacking chlorophyll and conductive tissue, and reproducing by spores.

Fusiform: A term describing a spike that is widest in the middle and tapers at tip and base.

G: Abbreviation for guanine (2-amino-6-hydroxy-purine), a purine base characteristic component of nucleic acid.

Gamete: Cell of meiotic origin specialized for fertilization.

Gamete selection: Involves crossing a good inbred line with a random sample of pollen from an open pollinated variety followed by selfing of individual F_1 plants.

Gametophytic self-incompatibility (GSI): In this incompatibility phenotype of the pollen is determined by its haploid genome.

Gamma rays: Short, wavelength electromagnetic radiation emitted from nuclei of radioactive isotopes.

Geitonogamy: Identical to autogamy.

Gel electrophoresis: An electrophoretic separation method in a gel matrix usually made of agarose, starch, or polyacrylamide. A large number of gel electrophoretic techniques are available to separate DNA molecules, including crossed field gel electrophoresis, denaturing gradient gel electrophoresis, field inversion gel electrophoresis, pulsed field get electrophoresis and temperature sweep gel electrophoresis.

Gene: The unit of inheritance. Genes are located at fixed loci in chromosomes and can exist in a series of alternative forms called alleles. It is unit of inheritance located mainly in the chromosomes at fixed loci and contains genetic information that determines characteristics of an organism.

Genebank: An establishment that preserves and maintains seeds, plants, and DNA of cultivated, wild and endangered species.

Gene centre: Centre of origin of a given crop plant.

Gene dispersal: Spread of genes within population through pollen and seeds. The term is often used indiscriminately for gene-flow.

Gene editing: This process typically looks to improve a beneficial trait within an organism, or to remove an undesirable trait. Gene editing tools like CRISPR are already helping researchers to make improvement within plant DNA. Here it is important to note that although plant breeding is a form of genetic engineering, it is not the same as genetic modification, or GM.

Gene expression: The phenotypic consequence of transcription of one or more specific genes.

Gene flow: Spread of genes among populations through pollen or seed (syn. migration).

Gene frequency: Proportion in which alternative alleles of a gene occur in a population.

Gene interaction: Modification of gene action by a non-allelic gene or genes.

Gene isolation: The identification of specific genes (each encoding a specific protein) responsible for a given trait and the isolation of those genes, usually via a bacterial plasmid or phage vector cloning system.

Gene pool: The reservoir of different genes of a certain plant species (lower and higher taxa) available for crossing and selection; divided based on hybridization into primary, secondary, and tertiary gene pools.

Gene pyramiding: Assembling many alleles with similar effects from different loci in a single genotype. This is achieved by marker based recurrent selection (MBRS).

Gene probe: A specific single stranded DNA sequence, often labeled with radioactive phosphorous (32p) to help the detection of a gene through the complementary sequence to which the probe will bind. Hybridization is normally detected by autoradiography.

Gene sanctuary: An area within the centres of diversity protected for human interference.

Gene silencing: Absence of expression of a gene which is present in the genome.

Gene therapy: The treatment of inherited diseases by introducing wild type copies of the defective gene causing the disorder into the cells of affected individuals. If reproductive cells are modified; the process is called germ-line or heritable gene therapy. If cells other than reproductive cells are modified; the procedure is called somatic cell or nonheritable gene therapy.

Gene transfer (transformation): The coding regions of isolated genes, along with appropriate regulatory signals, are moved into a recipient organism/ crop plant on a vector in such a way that the genes are integrated into the chromosome and expressed.

General resistance: Non-specific host plant resistance.

Genetic advance: The expected gain in the mean of the population for a particular quantitative character by one generation of selection of a specified per cent of the highest ranking plants.

Genetic code: The conversion cipher that allows the interpretation of nucleotide triplet (codons) to their matching amino acid.

Genetic diversity: Variations in genotypes resulting from hybridization and selection under the law of the survival of the fittest over many millennia.

Genetic drift: Changes in gene frequencies in a finite population due to stochastic effects such as random sampling of gametes; opposed to deterministic changes in gene frequencies caused by selection.

Genetic engineering: Genetic manipulations that use recombinant DNA methods (gene splicing) to change the genetic make-up of an organism.

Genetic equilibrium: In a random mating population, the stage in which genotype frequencies do not change from one generation to another.

Genetic erosion: Gradual disappearance of various forms of a cultivated species and of its wild relatives. The loss of genetic information that occurs when highly adaptable cultivars are developed and these threaten the survival of their more locally adapted ancestors, which form the genetic base of the crop.

Genetic fingerprinting: A method that uses DNA technologies to establish genetic relationships between close or distant relatives.

Genetic load: The sum total of deleterious (harmful) alleles present in a random mating or Mendelian population.

Genetic map: An experimentally determined diagram which shows the relative position of genes on chromosomes.

Genetic marker: Marker which is based on molecular or biochemical properties.

Genetic material: Single or double stranded nucleic acid that serves as a template of its own replication or for the synthesis of structural or mRNA.

Genetic resources: Germplasm containing potentially useful characteristics of plants, animals and other organisms.

Genetic vulnerability: Refers to pest susceptibility when large geographic regions are planted with the same crop cultivar or with cultivars having the same genes or cytoplamic factors.

Genetics: The science of heredity and variation.

Genome: A set of chromosome corresponding to the haploid set of a species.

Genomic DNA library: The collection of clones containing the genomic DNA sequences of an organism.

Genomic selection: Selection based on genomic breeding values that are calculated as the sum of the effects of dense genetic markers, or haplotypes of these markers, across the entire genome, thereby potentially capturing all QTL that contribute to variation in a trait and similarly to multiple traits.

Genomics: The field of study that seeks to understand the structure and function of all genes in an organism based on entire DNA sequence with extensive reliance on powerful computer technologies.

Genotype ratio: The proportions of the different genotypes in a particular progeny.

Genotype: The entire genetic constitution of an organism. It is the description of the genes contained by a particular organism for the character or the characters –the entire genetic constitution.

Genotyping: Rather than just being able to see the results of breeding through phenotyping, one can see what happens to the structure of DNA and know why these changes occur in the plants at a genetic level. This means that scientists are now using technology to identify individual genes within plants, giving them a deep understanding of exactly what clusters of DNA are responsible for certain traits and characteristics. This gives scientists an unprecedented ability to develop seed varieties for specific environments and markets.

Genus (pl. genera): a taxonomic category that includes groups of closely related species.

Germplasm: The sum total of the hereditary materials in a species.

Germplasm enhancement: A process of improving a germplasm accession by breeding while retaining the important genetic contributions of the accession.

Germplasm resources: Any materials of potential use in breeding programmes that are not themselves of immediate value as cultivars of the crop under consideration. It includes present and obsolete cultivars, landraces, breeding lines, wild or weed forms, and crop relatives.

Gibberellins: A group of plant hormones that dramatically stimulate plant growth by cell division, cell elongation or both.

Gigas: Exhibiting large sized characteristics.

Glabrous: Smooth, especially not pubescent/hairy.

Glassine bag: A glazed, semi-transparent paper that withstands outdoor conditions and is used to protect emasculated flowers from cross-pollination.

Glume: Two bracts found at the base of a grass spikelet.

Gluten: A mixture of proteins that results in dough elasticity and cohesiveness and is found most abundantly in wheat and rye.

GMO: Genetically modified organism.

Golden rice: A transgenic line of rice that is enhanced for beta carotene (pro-vitamin A). GM rice that produces beta carotene in the endosperm and shows yellow colour of the grain that is visible after milling and polishing from which generic name 'Golden Rice' is derived. This technology/ product was developed by I. Potrycus and P. Bayer and co-workers and was funded by Rockefeller Foundation, The Swish Federal Institute of Technology, the EU and the Swish Federal Office for Education and Science.

Gossypol: A polyphenolic compound present in darkly pigmented glands that occur throughout the cotton plant.

Green house effect: An increase in carbon dioxide in the earth's atmosphere that has a warming effect, as it restricts re-radiation of sunlight to the outer atmosphere.

Green revolution: The development and successful introduction of high yielding semi-dwarf wheat and rice cultivars in Asia and Latin America.

Guanine: A purine base found in DNA and RNA.

Guide RNAs: RNA molecules that contain sequences that functions as templates during RNA editing.

Gymnosperm: A seed plant with seeds not enclosed in an ovary wall.

Gynoecium: Collective name for stigma, style and ovary.

Gynophore: An elongation of floral axis between the stamens and the carpel, forming a stalk that elevates the gynoecium.

Habit: The general appearances of a plant.

Hair: A slender outgrowth of the epidermis, common on certain leaf structures.

Half-sib: Offspring possessing one parent (usually the mother) in common.

Haploid: A cell or organism with the gametic chromosome number (n).

Haploid breeding: A procedure in plant breeding to obtain haploids through anther culture followed by chromosome doubling with colchicine to produce double haploids (DH).

Hemiploid: An individual with haploid chromosome number.

Hardy-Weinberg equilibrium: The allele or gene frequencies reached in one generation of random mating population given by the expansion $(p + q)^2 = (p^2 + 2\,pq + q^2)$ where p and q represent the frequency of allele 'A' and 'a' respectively. It is a mathematical relationship that allows the frequencies of genotypes in a population to be predicted from their constituent allele frequencies, a consequence of random mating.

Harvest index: Grain weight as a percentage of total above- ground dry weight at maturity.

Hemizygous: The condition in which only one allele of a pair is present, as in sex linkage as a result of deletion.

Heredity: The transmission of characteristics from parents to their offspring. The process that brings about biological similarity and dissimilarity between parents and progeny.

Heritability: The proportion of observed variability which is due to heredity, the remainder being due to environmental causes. more strictly, the proportion of observed variability due to the additive effects of genes (narrow sense heritability).

Hermaphroditism: Reproductive organs of both sexes present in the same individual or in the same flower in higher plants.

Heteroalleles: Mutations that are functionally allelic but structurally non-allelic, mutations at different sites in a gene.

Heterocaryosis: The presence of two or more genetically different nuclei within single cells of a mycelium.

Heterochromatin: Portion of genomic DNA that is highly stained because of high condensation. The heterochromatin is transcriptionally inactive and often contains large amounts of highly repeated DNA (satellite DNA). It is often found around centromere.

Heteroduplex: A double stranded nucleic acid containing one or more mismatched base pairs.

Heterogamy: Separation of anthers and stigma in space within a flower in such a manner that self-pollination is avoided.

Heterosis: Hybrid vigour such that an F_1 hybrid falls outside the range of the parents with respect to some character or characters.

Heterostyly: Occurrence of style and stamens of different lengths in flowers from different plants on a single species.

Heterozygous: Having unlike alleles at one or more corresponding loci (opposite of homozygous).

Hexaploid: Having six sets of chromosomes; chromosome number of 6x.

Hills: In corn refers to closely clustered groups of plants in contrast to row planting.

Homoalleles: mutations those are both functionally and structurally allelic.

Homoeologous: Having homology, as with homologous or partially homologus chromosomes originating from different genomes.

Homology of chromosome: Applied to whole chromosome or parts of chromosome which synapse or pair in meiotic prophase.

Homology: The extent of identity between nucleotide or amino acid sequences.

Homozygous: Having like alleles at corresponding loci on homologous chromosomes.

Hooded: Refers to awns on glumes that terminate in various wings like structures.

Horizontal resistance: Resistance governed by polygenes and non-specific to particular race.

Hull: The outer covering of a fruit or seed in cereal crops; hulls may be removed freely as in wheat or adhere as in barley.

Hybrid: The product of a cross between genetically unlike parents.

Hybrid variety: Commercially exploitable F_1 hybrids between two different parental lines.

Hybridization: (1) The crossing of individuals of unlike genetic constitution; (2) a method of breeding new varieties that utilize crossing to obtain genetic recombination.

Hypersensitive sites: Regions in the DNA that are highly susceptible to digestion with endonucleases.

Hypogeal emergence: The elongation of epicotyls and the cotyledons remaining below ground, common in monocots.

Hypomorph: A mutation that reduces but does not completely abolish gene expression.

I_1, I_2, I_3: Symbols used to designate first, second, third etc. inbred generations (= S_1, S_2, ...).

Ideoptype: An ideal plant form formulated to assist in reaching selection goals.

Ideotype breeding: A method of breeding to enhance genetic yield potential base on modifying individual traits where the breeding goal (phenotype) for each trait is specified.

Immune: Completely free from attack by a given pathogen.

Imperfect flower: A floret lacking either stamens (pistillate) or a pistil (staminate).

Improved seed: Seed of an improved variety having a very high genetic and physical purity and germination.

Inbreeding: Mating of closely related plants or animals, the mating of individuals more closely related than individuals mating at random.

Inbreeding coefficient: A quantitative measure of the intensity of inbreeding.

Inbred-line: A line produced by continued inbreeding. In plant breeding, nearly homozygous line is usually originating by continued self-pollination, accompanied by selection.

Inbred variety cross: The F_1 cross of an inbred line with a variety.

***In vitro*:** Conducted outside a living organism, in contrast to in vivo, where the process happens within a living organism.

***In vivo*:** From the Latin meaning within the living organism.

Incompatibility: Failure to obtain fertilization and seed formation after self-pollination, usually due to failure of pollen tube to penetrate the stigma, or reduced growth of the pollen tube in the stylar tissue.

Indehiscent: A fruit that does not open by sutures, pores, or caps, the seeds being released by the rotting away of the fruit wall.

Independent assortment: The chance distribution of two or more pairs of segregating genes to the gametes.

Indeterminate: A type of plant growth in which terminal bud remains vegetative resulting in overlapping of vegetative and reproductive stages.

Indexing: The process used to test vegetatively reproduced planting stock for freedom from virus diseases.

Indicator row: Susceptible check sown together with test entries to assess the disease pressure.

Indigenous: An organism originated naturally in a particular area.

Indirect selection: Improvement of a character achieved by selection for another related trait.

Induced mutation: Induction of mutation by mutagens.

Inflorescence: (1) A flower cluster; (2) the arrangement and mode of development of the flowers on a flower axis.

Inherit: Receive from one predecessor. In organism, chromosomes and genes are transmitted from one generation to the next.

Inoculate: To place inoculum where it will produce an infectious disease.

Inoculum: Spores, bacteria, or fragment of mycelium of pathogens that can infest plants or soil.

Interaction: The tendency of one factor to have different effects at different levels of another factor.

Inter-crossing: Random mating among cross-pollinated plants.

Interference: The effect of one crossover influencing the probability that another will occur in the immediate vicinity.

Internode: The portion of a stem or other structure between two nodes.

Inter-specific cross: The cross between two parents from different species.

Introgression: Transfer of a few genes from one species into the full diploid chromosome complement of another species.

Intron: Intervening sequence in genes that is transcribed but not translated.

Inversion: A rearrangement of a chromosome segment so that its genes are in reversed linear order.

Involucre: Protective group of bracts enveloping sexual parts of a young inflorescence.

Ion: An atom that has lost or gained an electron in comparison to the number of electrons normally present.

Ionization: Loss or gain of an electron by an atom.

Ionizing radiation: The portion of the electromagnetic spectrum that results in the production of positive and negative charges (ion pairs) in molecules. X-ray and gamma-rays are examples of ionizing radiation.

Irradiation: Exposure of plants parts to X-rays or other radiations to increase mutation rate.

Irregular flower: The perianth constituted in such a way that it is not possible to divide it into two equal halves, as in canna flower.

Isoalleles: Alleles indistinguishable except by special tests.

Isogenic lines: Lines differing from each other genetically at one locus only i.e. lines identical in all traits but one.

Isolation: The separation of one group from another so that mating between or among groups is prevented.

Isozyme (isoenzyme, allozyme): One of multiple forms (usually electrophoretic) of a single enzyme resulting from either allelic variation within one polypeptide or multimeric associations of variant.

Juvenile phase: An early stage in the development of a plant when tillers develop, internodes elongate and the plant is insensitive to external stimuli.

Kneeing ability: A positive geotropic response whereby the floating leaves and stem of the deep-water or floating rice become erect.

Karyogamy: The fusion of two gametic nuclei.

Karyotype: The pattern and shape of the complete set of chromosomes belonging to an organism.

Key: An abbreviated statement of contrasting diagnostic characters.

Kilobase (kb): One thousand base pair.

Kinetochore: Spindle attachment. A localized region in each chromosome to which the spindle fibre appears to be attached and which seems to determine movement of the chromosomes during mitosis and meiosis.

Kit: A set of reagents necessary to perform an experimental protocol.

Labeling. The introduction of radioactive or non-radioactive atoms or chemical groups in biological molecules (e.g. DNA. RNA, proteins) for their identification.

Lambda: The temperate bacteriophage of *Escherichia coli* that contains a double-stranded DNA genome of 49 kb packaged into an icosahedral head.

Lamina: The blade or expanded part of a leaf, petal, etc.

Lanceolate: Having a slender or spear-headed shape.

Landraces: Traditional cultivars with sufficient genetic integrity to be morphologically identifiable and differing in adaptation and to cultural practices, but genetically variable.

Lateral: Belonging to or borne on the sides.

LD_{50}: The dose of any agent which kills 50 per cent of the treated individuals.

Leader sequence: The segment of an mRNA molecule from the 5 terminus to the translation initiation codon.

Leavened: A dough product that has undergone fermentation by addition of yeast or baking powder to lighten the dough.

Lethal genes: Usually recessive in nature lethal genes kill each and every individual that carries them in homozygous state.

Ligase: An enzyme used to couple termini of one or more double stranded DNA molecules together.

Ligation: The formation of a phiosphodiester bond between neighbouring nucleotides exposing a 5 phosphate and 3 hydroxyl group.

Ligule: A projection from the summit of the sheath.

Line breeding: A system of breeding in which a number of genotypes which have been progeny tested in respect to some characters or group of characters are composited to form a variety.

Linear: Long and narrow, with parallel margins.

Linkage analysis: Estimation of cross-over frequency or recombination between DNA sequences, as the means to establish the relative position of a locus (e.g. gene, sequence tag) in a chromosome.

Linkage drag: When transferring a single gene from a donor into the genetic background of a recurrent parent by repeated back-crossing, genetic linkage will cause fragments of donor genome surrounding the target gene to be 'dragged' along and this is called 'linkage drag'.

Lint: Long fibres of cotton separated from seed in ginning.

Locus: The position occupied by a gene in a chromosome.

Lod score: The logarithm of the ratio of odds for pedigree data, where the odds are calculated under the assumption of linkage with a specified frequency of recombination and under the assumption of no linkage, that is, 50 per cent recombination.

Lodging: The bending or breaking over of a plant before harvest.

Long day plants: Plants that require a short night or dark period for reproductive development. Although it is the length of night that influences the reproductive development, the original classification of long-day, short day and day neutral plants persists.

M_1, M_2, M_3: Symbols used to designate first, second, third generations after treatment with a mutagenic agent.

MABC-Molecular accelerated backcrossing: A molecular marker is used for indirect selection during backcrossing of the gene/QTL of interest into elite material.

Maintainer line (B-line): Line used for maintaining a cytoplasmic male sterile line. It has the same nuclear genotype as the male sterile line (A-line).

Male sterility: Absence or non-functioning of pollen in plants. It may be genetic, cytoplasmic or cytoplasmic-genetic.

Map: The ordered arrangement of genes or molecular markers of an organism, indicating the position and distance between the markers and loci. Most maps are genetic maps based on the recombination percentage. Some maps are cytological maps based on the arrangement of chromosomal regions. There are physical maps based on the amount of DNA between markers and loci.

Mapping: Assigning markers, genes, and/or QTL in order, indicating the relative distances among them, and assigning them to their linkage groups on the basis of their recombination values from all pair-wise combinations.

Map unit: Defined as one centi-Morgan.

Marker: A distinguishing feature that can be used to identify a particular part or region of genome, chromosome or genetic linkage map, also any molecule (e. g. DNA, RNA, protein) of known size, used as a standard to calibrate the electrophoresis or chromatographic separation of molecules.

Marker assisted selection (MAS): Increased breeding selection efficiency achieved using DNA markers that allow earlier selection of positive traits and elimination of negative traits in a plant population. The breeder can potentially reduce the population size more rapidly in selecting for qualitative or quantitative traits.

Mass selection: A form of selection in which individual plants with desirable phenotypes are selected from a large population of plants grown in a field and the next generation propagated from the aggregate of their seeds.

Mass pedigree method: A system of breeding in which a population is propagated in mass until conditions favourable for selection occur, after which pedigree selection is practiced.

Maternal effect: Trait controlled by a gene of the mother but expressed in the progeny.

Maternal inheritance: Inheritance controlled by extra chromosomal (that is, cytoplasmic) factors that are transmitted through the egg.

Mating system: Any of a number of schemes by which individuals are assorted in pairs leading to sexual reproduction. The different types of mating systems are random mating, genetic assortative mating, genetic disassortative mating, phenotypic assortative mating and phenotypic disassortative mating.

Mb (mega base): One million base pairs.

Megagametogenesis: Production of female gametophyte (embryo sac) from megaspore (through mitosis).

Megaspore: One of the four haploid spores originating from the meiotic divisions of the diploid megaspores.

Megaspore mother cell: Diploid cell in ovary that gives rise, through meiosis, to four haploid megaspores.

Meiosis: Two successive nuclear divisions, in the course of which the diploid chromosome number is reduced to the haploid.

Melting: It is the dissociation of complementary strands of double stranded nucleic acids, such as homoduplex (dsDNA), or heteroduplex (DNA-RNA) molecules. Melting can be accomplished in vitro by heating or alkali treatment or in vivo by specific binding protein involved in replication and translation.

Mendelian population: A natural interbreeding unit of sexually reproducing plants or animals sharing a common gene pool.

Meristem culture: Cultivation of apical meristems, particularly shoot apical meristem, for production of shoots and plantlets.

Meristem regions: Areas in the plant where cells are rapidly dividing.

Metaphase: The stage of cell division at which the chromosomes lie on the metaphase plate.

Metaxenia: Influence of pollen on maternal tissues of the fruit. Tissues outside the embryo sac are influenced by the pollen source.

Methylation: Enzymatic addition of a methyl (CH_3) group from a methyl donor to DNA that causes inactivation of that region. Usually CpG nucleotide pairs are target for this addition.

Microgametogenesis: Production of male gametophyte from microspore mother cell (through mitosis).

Microprojectile bombardment: A process for transforming plant cells by shooting DNA coated tungsten or gold particles into the cells.

Micropropagation: Asexual in vitro production of plants using meristems or other plant parts as starting tissue, which provides an opportunity to produce plants that are disease free and that can be used for cell manipulation.

Microsatellite (short tandem repeats-STR, simple sequence repeats-SSR): One of many DNA sequences dispersed throughout fungal, plant and animal genomes, composed of short (2-10bp) sequences that are repeated in tandem and are usually highly variable and useful for DNA profiling.

Microspore: One of the four haploid spores originating from the meiotic division of the microspore mother cell in the anther and which gives rise to the pollen grain.

Microspore mother cell: Diploid cell in the anther which gives rise, through meiosis, to four haploid microspores.

Microsporogenesis: Production of pollen grains (microsproes) from microspore mother cell (through meiosis).

Missense mutation: A mutation that changes a codon specifying an amino acid to a codon specifying a different amino acid.

Mitochondria: Rod like or spherical bodies found in the cytoplasm of all cells and containing enzymes for respiration.

Mitosis: The process by which the nucleus is divided into two daughter nuclei with equivalent chromosome complements, usually accompanied by division of the cell containing the nucleus.

Model: A computerized simulation of scientific events in crop development based on a series of logically assembled, justifiable beliefs about a system.

Modifying genes: Genes that affect the expression of a non-allelic gene or genes. The gene may be enhancer (intensifies the expression) or reducer (decreases the expression of other gene).

Molecular marker: A molecular signpost used in eukaryotic gene isolation, usually an RFLP probe or a sequence tagged site of DNA amplification of hybridization.

Monocotyledonous: Having a single cotyledon.

Monoecious: Staminate and pistillate flowers borne separately in the same plant.

Monoecy: State of being monoecious.

Monohybrid: Heterozygous with respect to one gene.

Monoploid: An organism with the basic (n) chromosome number.

Monosomic: An organism lacking one chromosome of the diploid complement (2n-1 chromosomes).

Morphology: The form, structure, and development of plants.

mRNA: A single stranded RNA molecule that is product of the transcription of DNA and the information carrier for translation into protein.

Multiline variety: A blend or mixture of cultivars of one species that are genetically very similar but differ in specificity of resistance to strains of a disease.

Multiple genes: Two or more genes at different loci, which produce complementary of cumulative effects on a single quantitative genetic trait.

Multivalent: The structure formed by association of more than two homologous chromosomes as a consequence of synapsis during meiosis.

Mutagens: Substances that are known to induce genetic changes or mutations.

Mutation: A sudden heritable change in a gene or in chromosome structure.

Mutation breeding: The use of mutagenic agents to increase the frequency of mutant plants useful in the breeding of improved varieties.

Mycelium: A mass of thread like filaments (hyphae) of a fungus comprising the main body.

Mycoplasma: Microscopic, unicellular organisms similar to but smaller than bacteria and lacking rigid cell walls.

N: Abbreviation for any one base in a nucleic acid molecule (e.g. N stands for G, C, A or T in DNA).

Natural crossing: It occurs between individuals within a population.

Natural selection: Selection due to natural forces, i.e., environment.

Near-isogenic line (NIL): A line used frequently to define molecular regions of a genome that causes a phenotype of value in agronomy and plant breeding.

Net assimilation rate: Total rate of photosynthesis minus respiration.

Nick: A break of a phosphodiester bond in only one of two strands of a nucleic acid duplex.

Nicking: Synchronization of the receptivity of the male-sterile plant to the maximum pollen load of the pollinator for cross-pollination in hybrid seed production.

Nodal bud: The lateral shoot bud located within the root ring at the node.

Non-recurrent parent: Parent not involved again in backcross, also donor parent.

Northern transfer (blotting, hybridization): RNA detection method whereby RNA molecules separated by size are transferred directly to nitrocellulose filter or other membranes by electric or capillary force. Single stranded molecules can be immobilized to the membrane by baking, and are then hybridized to a labeled probe. The procedure is similar to Southern hybridization (DNA), and detects individual RNA species from a complex RNA.

Nubbin: A small, poorly developed corn ear.

Nucleic acid: A single or double stranded polynucleotide containing either deoxyribonucleotides (e.g. DNA) or ribonucleotides (e.g. RNA) linked by 3'-'5 phosphodiester bonds.

Nueleoside: A purine or pyrimidine base covalently linked to ribose or deoxyribose via N-glycosidic bond.

Nucleotide (nt): A purine or pyrimidine nucleoside that is esterified with one, two or three phosphate groups at the 5' C of ribose or deoxyribose. Nucleotides are components of DNA/RNA with base paired according to Chargaff rules (A=T, G=C in DNA and A=U, G=C in RNA).

Nulliplex: The condition in which a polyploid is recessive in all chromosomes in respect to a particular gene. Simplex denotes recessives at all loci except one, duplex two, triplex three, quadruplex four, etc.

Nullisome: An otherwise 2n plant that lacks both members of one specific pair of chromosomes, hence, with 2n-2 chromosomes.

Obconical: Inversely conical.

Oblong: More or less elliptic but with somewhat parallel sides.

Obovate: Reverse egg shaped, broadest above centre.

Obovoid: Egg shaped with the narrow end at the base.

Oligogenes: Genes having large individual effects producing distinct phenotypes.

Oligonucleotide: A short nucleic acid molecule containing up-to 100 deoxyribonucleotides (oligodeoxynucleotide) or ribonucleotides (oligoribonucleotides) or a mixture of both. These molecules are usually obtained synthetically.

Omes and Omics: These are various terms ending omes and omics. These include cytomics, epigenomics, genomics, immunomics, interactome, metabolomics, phenomics, proteiomics, secretome, transcriptomics, transgenomics, etc.

Ontogeny: The development of an organism from fertilization to maturity.

Open pollination: Natural cross-pollination.

Optimum leaf area: The leaf area at which dry matter production is maximum.

Outbreeding: Inducing cross fertilization in a normally self fertilized plant.

Outcross: A cross, usually natural, to a plant of different genotype.

Ovary: The enlarged basal portion of the pistil in which the seeds are borne.

Ovate: Egg shaped; having an outline like that of an egg, with the broader end basal.

Overdominance: The combined effect on a genetic trait of two alleles such that the heterozygote is more extreme than either homozygote (the effect, of Aa is greater than either AA or aa).

Ovule: The structure that bears the female gamete and becomes the seed after fertilization.

Oxalic acid: A bitter, poisonous acid.

Pachytene: The double-thread stage of meiosis.

Palea: The upper of the two bracts enclosing each floret in the grass spikelet.

Panicle: An open and branched inflorescence in which spikelets are borne on subdivided branches arising from a main axis.

Panmixia: Random mating without restriction (usually extended to include random mating under the restrictions of sex or incompatibility).

Parameter: A value or constant based on an entire population.

Parenchyma: Soft tissue of cells with un-thickened walls.

Parthenocarpy: The production of fruits without fertilization and, normally, without seeds.

Parthenogenesis: The development of an individual from a gamete without fertilization.

Partial dominance: Lack of complete dominance, the production of a hybrid intermediate between one of the parents and the mid-parental value.

Partial resistance: Resistance locus/QTL that do not provide a complete resistance irrespective of the mode of its inheritance. QTL are always partially expressed, but some monogenically inherited resistance genes also show partial expression, e.g., *Lr34*, *Yr36*, and *Pch1*. The term partial resistance, often used synonymously with quantitative resistance, is therefore ambiguous.

Patent: A limited property right granted by the government allowing investors to exclude others from making, using or selling their inventions without their approval for a specific time.

Pathogen: An organism capable of causing a disease.

Pathogenicity: The ability of an organism to cause a disease.

Pathosystem: Combination of a specific host and pathogen species or a complex of closely related pathogen species.

Pedicel: The stalk of a single flower.

Pedicellate: Borne on a pedicel.

Pedigree: A record of the ancestry of an individual, family, or strain.

Pedigree breeding: A system of breeding in which individual plants are selected in the segregating generations from a cross on the basis of their desirability judged individually and on the basis of a pedigree record.

Peduncle: Stalk of an inflorescence.

Penetrance: The frequency with which a gene produces a recognizable effect in individuals which carry it.

Pentaploid: Having five sets (genomes) of chromosomes; chromosome number of 5x.

Perennial: The plants continue to grow through years (2 or more), flower and fruit repeatedly.

Perfect flower: Flower possessing both stamens and pistils.

Perfect marker: A marker without recombination to the gene of interest, ideally drawn directly from the gene sequence.

Phenetic (pattern, diversity) analysis: Study of patterns of overall resemblance and difference among organisms based on many heritable characteristics. The patterns can reveal groupings with different degrees of variation within and between groups that reflect the relationship between organisms at a given time.

Phenomics: Phenomics is a field of study concerned with characterization of phenotypes which are the characteristics of the organisms that arise via the interaction of the gene with the environment.

Phenotype: Physical or external appearances of an organism as contrasted with its genetic constitution (genotype).

Phenotypic ratio: The proportions of the different phenotypes in a particular progeny.

Phenotypic value: The numerical description of a phenotype with respect to a particular quantitative character measured in metric units.

Phenotyping: Phenotyping means assessing a plant's expressed traits and then selecting the desired plants and seeds. In practical terms this means visually identifying differences between plants, for example, selecting for desirable colour, sizes, or number of grains, fruits, etc.

Phloem: Conducting tissue within a plant to transport the products of photosynthesis to other plant parts.

Photoperiodism: Reproductive response to the relative lengths of light and dark periods in a day.

Photorespiration: A form of respiration stimulated by light, found in C_3 plants.

Photosynthesis: The process of transforming carbon dioxide and water into carbohydrate (food) the basis of all crop yields and essence of agriculture.

Phylogenetic tree: A relational graphic representation reflecting ancestral or evolutionary relationship in a group of organisms.

Phylogeny: The evolutionary development of a species or genus.

Physiological maturity: The state in grain development at which no further increase in dry weight occurs.

Physiological race: Pathogens of the same species with similar or identical morphology but differing pathogenic capabilities.

Physiology: The study of process taking place in living organisms.

Pistil: The seed bearing organ in the flower, composed of the ovary, the style, and the stigma.

Pistillate flower: A flower bearing pistils but no stamens.

Pith: The spongy centre of an exogenous stem, chiefly consisting of parenchyma.

Placenta: Any part of the interior of the ovary which bears ovules.

Plant architecture: The number, size shape, structure, arrangement, and display of particular plant parts.

Plant breeders' rights: Legal rights of ownership granted by government to the developer of a new cultivar. Rights refer to sale of the cultivar as seed, not to other uses, such as food or feed.

Plant breeding: The branch of biology concerned with an organized attempt to produce progressively better adapted plants. This is also defined as the art and science of changing the genotype of plants in the desired direction.

Plant introduction: (1) Transport of a collection of seeds, plants, or vegetative propagating materials from one ecological area into another; (2) collection of seeds, plants, or vegetative propagating materials that have been transported from one area into another.

Plant variety protection (PVP): Legislation that exists in many countries that protects the commercial use of varieties.

Plasmid: A relatively small, autonomous circular piece of DNA not part of chromosomes, most common in bacteria.

Plasmogamy: Fusion of cytoplasm of two or more cells.

Pleiotropy: Phenomenon of a single gene affecting two or more different characters.

Plumule: The growing point of the embryo.

Point mutation: A change in single base pair (gene mutation).

Polar nuclei: Two centrally located nuclei in the embryo sac that unite with the second sperm in a triple fusion. In certain seeds the product of this triple fusion develops into the endosperm.

Pollen grain: The male gametophyte, originating from a microscope.

Pollen tube: A tube developing from the germinating pollen grain. The sperm cells pass through the pollen tube to reach the ovule.

Pollination: Transfer of pollen from anther to the stigma. Self-pollination is the transfer of pollen from an anther to the stigma of the same flower or another flower on the same plant, or within a clone. Cross pollination is the transfer of pollen from an anther on one plant to a stigma in a flower on a different plant.

Polygene: A gene that individually exerts too slight effects but together with several or many genes controls quantitative character.

Polyacrylamide gel: An isolable matrix of acrylamide monomers cross linked with N, N- methylene bisacrylamide by the presence of catalysts (e.g. TEMED) that can be made to harbor varying pore sizes.

Polycross: An isolated group of plants or clones arranged in some fashion to facilitate random inter-pollination.

Polygenes: Multiple genes i.e. numerous genes that influence a single character, each having a small effect.

Polygenic inheritance: Inheritance of quantitative characters due to action of polygenes.

Polyhaploid: Haploid plant produced from a polyploid plant; contains a multiple of the basic (x) chromosome number for that species.

Polymerase: An enzyme that catalyses the assembly of nucleotides into RNA (RNA polymerase) or DNA (DNA polymerase).

Polymerase chain reaction (PCR): An in vitro amplification method capable of increasing the mass (number) of a specific DNA region (fragment). The method uses two synthetic oligonucleotide primers (15-30 nucleotides long) that specifically hybridize to opposite DNA strands flanking the region to be amplified. A series of temperature cycles involving DNA denaturation, primer annealing, and extension of the annealed primers by the activity of a thermo- stable DNA polymerase, specifically and exponentially (at least during first many cycles) amplify the target DNA region many million fold to produce an amplified fragment with its termini defined by the 5 end of each primer. The term polymerase chain reaction has been used in a broad sense to define any amplification technique driven by primers, DNA polymerase and temperature cycling regime.

Polymorphism: The occurrence together in the same population of two or more distinct forms at frequencies too great to be explained by recurrent mutation.

Polyploid: An organism with more than two sets (genomes) of chromosomes in its body cells.

Population: In genetics, a community of individuals which share a common gene pool. In statistics, hypothetical and infinitely large series of potential observations among which observations actually made to constitute a sample.

Population genetics: The branch of genetics that deals with frequencies of alleles and genotypes in breeding populations.

Prepotency: The capacity of a parent to impress characteristics on its offspring so they are more alike than usual.

Primer: A short oligonucleotide complementary to a larger nucleic acid molecule which serves as a substrate of a DNA or RNA dependent DNA polymerase.

Probability: The frequency of occurrence of an event.

Probe: A defined nucleic acid sequence which is labeled and used to detect homologous sequences in DNA or RNA by hybridization.

Progeny testing: A test of the value of a genotype based on the performance of its offspring produced in some definite system of mating.

Prokaryote: Organism characterized by the absence of major organelles (nucleus, plastids). It includes primarily bacteria, etc.

Promoter: A nucleotide sequence to which RNA polymerase binds and initiates transcription, also a chemical substance that enhances the transformation of benign cells into cancerous cells.

Protandry: Maturation of anthers before pistils.

Protogyny: Reverse of protandry i.e., the female reproductive structures mature before the male structures.

Protoplast: A plant cell devoid of a cell wall.

Pseudogamy: In plants where an embryo develops form a diploid or haploid egg cell with a sperm penetrated but without fertilization and egg cell nucleus remains functional.

Pseudogenes: These are inactive genes but are stable components of the genome derived by mutation of an ancestral active gene.

Pubescent: Covered with hairs, especially if short and soft.

Pure-line: A strain homozygous at all loci, ordinarily by successive self-fertilization in plant breeding.

Pure-line selection: Isolation of pure lines from a mixture of pure lines.

Purine: A double ring nitrogen containing base present in nucleic acids, adenine and guanine are the two purines present in most DNA and RNA molecules.

Pyramid: To incorporate several genes for a specific trait into one variety.

Pyramiding resistance genes: Putting all the resistance genes into one genetic background.

Pyrimidine: A single ring nitrogen containing base present in nucleic acids, cytosine and thymine are commonly present in DNA whereas uracil replaces thymine in RNA.

Quadrivalent: A union of four homologous chromosomes at meiosis.

Qualitative character: A character in which variation is discontinuous.

Qualitative genetics: The inheritance of those differences among individuals that are characterized by discrete classes.

Qualitative resistance: Race-specific, monogenically inherited resistance, also named vertical resistance or hypersensitivity resistance although not all resistance genes show this phenotype, e.g., *Lr34*.

Quantitative inheritance: Inheritance of measurable traits (height, weight, colour intensity) that depend on the cumulative action of many genes, each producing a small effect on the phenotype.

Quantitative resistance: Resistance inherited by several genes with minor effects, usually non-race-specific and prone to nongenetic interactions, also named horizontal resistance.

Quantitative trait loci (QTL): A term given to genomic region which controls a phenotype by cooperative interaction of several genes (e.g. cyst nematode resistance in soybean).

Quantum speciation: The formation of a new species in one or few generations by selection and genetic drift.

Quarantine: Isolation of an organism for observation on weeds, diseases and pests and for preventing their spread.

Race: A genetically distinct group within a species.

Rachilla: A secondary axis or little rachis of a spikelet.

Rachis: The central axis of an inflorescence of a cereal crop plant.

R-line: The pollen parent line, containing fertility restorer genes, crosses with A line in the production of hybrids seeds.

R_0, R_1, R_2: Plant regenerated from cell culture irrespective of tissue of origin, are referred to as R or R_0 plants. The self fertilized progeny of R plants are referred to as R_1 plants.

Random amplified polymorphic DNA (RAPD) analysis: A widely used arbitrarily amplified DNA technique that uses one primer of typically 9 or 10 nucleotides in length, low annealing temperatures and low primer to template ratio (less than1). The amplification products are generally visualized by agarose gel electrophoresis and ethidium bromide staining, and fingerprints are simple and especially useful for genetic mapping.

Random: Arrived at by chance without discrimination.

Randomization: Process of making assignments at random.

Random mating: A system of mating in which an individual has equal chance of mating with every other individual of the same population.

Recessive: The member of an allelic pair which is not expressed when the other (dominant) member occupies the homologous locus.

Reciprocal crosses: Crosses in which the sources of male and female gametes are reversed.

Recognition site (sequence): Nucleic acid sequence to which a specific protein (e.g. restriction nuclease) can be bound.

Recombinant DNA: The insertion of DNA containing genetic information into an organism or cell using in vitro techniques.

Recombinant inbred line (RIL): The product of an initial cross between two parental lines and subsequent selfing of the F_2 individuals generally used for mapping purposes.

Recombination: Formation of new combinations of genes as a result of segregation in crosses between genetically different parents, also the rearrangement of linked genes due to crossing over.

Recumbent: Curved downward or backward.

Recurrent parent: The parent to which successive backcrosses are made in backcross breeding.

Recurrent selection: A method of breeding designed to concentrate favourable genes scattered among a number of individuals by selecting in each generations among the progeny produced by mating *inter se* of the selected individuals (or their selfed progeny) of the previous generation.

Reduction division: A nuclear division in which the chromosomes are reduced from the diploid to the haploid number.

Regeneration: The creation of complete plant from a single plant cell or small piece of plant tissue through in vitro culture.

Regression coefficient: A numerical measure of the rate of change of the dependent on the independent variable.

Replications: Repetition of treatments in experiments that allows for statistical analysis.

Reproduction: Production of a new generation of individuals (progeny) by sexual or asexual means.

Resistant: Characteristic of a host plant such that it is capable of suppressing or retarding the development of a pathogen or other injurious factor.

Restorer gene: A gene, usually dominant, that effectively overcomes the effect of male sterile cytoplasm on male sterility, i.e. produces functional male gametes even in the presence of the male sterile cytoplasm.

Restriction endonuclease: A bacterial enzyme that recognizes specific sites (recognition sequences) in DNA and cuts (restricts) the molecules into (restriction) fragments.

Restriction map: The linear arrangement of restriction sites on a nucleic acid.

Reverse transcriptase (RTase): An RNA dependent DNA polymerase enzyme capable of synthesis of a double stranded DNA from an RNA template.

Reversion: A change backward, as to an earlier condition.

RFLP: Restriction fragment length polymorphism (RFLP). It is the observable polymorphisms resulting from the restriction enzyme digestion of DNA from genetically district individuals. RFLPs are advantageous over morphological and isozyme markers primarily because their number is limited only by genome size and they are not environmentally or developmentally influenced.

Rhizome: An underground stem, usually horizontal and often elongated, distinguished from a root by the presence of nodes and internodes and sometimes scale like leaves and buds at the nodes.

Rib: A primary or prominent vein of a leaf.

RNA (ribonucleic acid): A single stranded polynucleotide of low molecular weight that contains a ribose backbone and pyrimidine and uracil instead of thymine. Constituent of messenger, transfer and ribosomal RNA and certain viruses (e.g. tobacco mosaic virus) and viroids. Some RNA molecules may have enzymatic activity (ribozymes). An intermediate chemical that translates genetic information into action.

Rogue: A variation from the standard type of a variety of strain, roguing or removal of undesirable individuals to purify the stock.

Root hair: Unicellular absorptive hairs on the young roots and rootlets.

Root: The descending axis of the plant, usually growing in the opposite direction from the stem, without nodes and internodes, mostly developing underground and absorbing moisture and nutrients from the soil.

S_1, S_2, S_3: Symbols for designating first, second, third etc., selfed generations from an ancestral plant (S_0).

Saline soil: A soil containing sufficient soluble salts to impair crop productivity.

Sample: A finite series of observations taken from a population.

Sampling error: Deviation of sample value from the true value owing to the limited size of the samples.

Scale: Any thin, scarious body, usually a degenerate leaf.

Scarification: A method for mechanically making slit on the seed coat usually for wild species seeds that facilitates imbibitions of water and germination. In a very hard seed coat, sulphuric acid is used for few seconds.

Scutellum: A shield-shaped body covering the embryo of a grass seed.

Scion: A portion of actively growing shoot of a plant is grafted onto a stock of another.

Secondary gene pool: Genes from secondary gene pool can be transferred to cultivated crops by primary gene pool with some difficulty compared to species of the primary gene pool.

Seed: A mature ovule with its normal covering. A seed consists of the seed coat, embryo, and in certain plants, an endosperm.

Seedling: A young plant grown from a seed.

Seed vigour: The overall ability of seed to perform well in the field.

Segregation: Separation of paternal from maternal chromosomes at meiosis and consequent separation of alleles.

Selection: (1) Any process, natural or artificial, that permits an increase in the proportion of certain genotypes or groups of genotypes in succeeding generations; (2) a plant, line, or strain that originated by a selection process.

Selection intensity: The percentage of the population selected.

Self fertility: Capability of producing seed upon self fertilization.

Self fertilization: Fusion of male and female gametes from the same individual.

Self incompatibility: Genetically controlled physiological hindrance to self fruitfulness.

Semigamy: Abnormal fertilization in which the male game fertilized the egg, but does not fuse with the egg nucleus.

Seminal: Arising from the seed.

Semolina: The end product of the ground up endosperm of durum wheat consisting of particles of certain size and uses to produce paste or paste products.

Sequence tagged site (STS): A small region of DNA on a chromosome detected by hybridization or amplification techniques and used as signpost for molecular gene mapping approaches.

Sessile: Without stalk.

Setae: (Singular, seta), bristles.

Sexual reproduction: Reproduction involving germ cells and union of gametes.

Shattering: The fortuitous loss of seed from a plant before harvest.

Sheath: A tubular envelope, as the lower part of the leaf in grasses.

Short day plant: A plant in which flowering is induced by daily exposure to less than 12 h of light.

Short term inbreds: A line derived by one or a few generations of inbreeding. Such a line is not homozygous or even nearly homozygous.

Sib-mating: The mating of sibs.

Sibs: Progeny of the same parents derived from different gametes. Half sibs, progeny with one parent in common.

Silencer: A DNA sequence that helps to reduce or shut off the expression of a nearby gene.

Single cross: A cross between two genotypes, usually two inbred lines, in plant breeding.

Single seed descent: Selection procedure in which F_2 plants and their progenies are advanced by single seed until genetic purity is virtually attained.

Sodic soil: A soil containing sufficient sodium to interfere with growth and productivity of crop plants.

Somaclonal variation: Variation occurring in plants derived from any form of cell culture.

Somatic cell: Non reproductive or vegetative tissue cell that is developed through mitosis and will not undergo meiosis.

Somatoplastic sterility: Collapse of zygote during embryonic stages due to disturbances in embryo endosperm relationship.

Southern hybridization (blotting, transfer): A gel blotting method whereby DNA fragments are separated by size and transferred to a nitrocellulose, nylon or other membrane or matrix by electrical or capillary action. Single stranded molecules are immobilized to the membrane and then hybridized to a labeled probe. The procedure has innumerable variants that detect individual DNA species from a complex mixture of DNA fragments.

Species: In classification, a subdivision of a genus. A group of closely related individuals descended from the same stock. A kind of plant or animal, its distinctness seen in morphological, anatomical, and cytological and chemical discontinuities presumably brought about by reproductive isolation in nature from all other organisms. These populations are capable of inter-breeding and producing viable fertile offspring.

Specific resistance: Host plant resistant to specific biotype of the pathogen; interaction of host plant gene-conditioning resistant reaction with a pathogen gene for pathogenicity conditioning avirulence.

Sperm: A male gamete.

Spores: Reproductive units of fungi.

Sporogenesis: Production of mega and micro spores from mega and microspore mother cell, respectively (through meiosis).

Sporophytic self-incompatibility (SSI): The incompatibility relationship of pollen is determined by the diploid plant producing it. It is controlled by single S locus with multiple alleles.

Stamen: The pollen bearing organ in the flower, composed on an anther and a filament.

Staminate flower: A flower bearing stamen but no pistil.

Standard deviation: A measure of variability, mathematically, the distance along the abscissa from the mean to the point of inflection of a normal curve.

Standard error: A statistic which is the estimated value of the standard deviation (a parameter).

Statistic: Estimate of a parameter made from a sample. Statistic is a sample what parameter is to population.

Sterile: Unproductive, as a flower without pistil or a stamen without an anther; infertile.

Sterility: Failure to complete fertilization and produce seed as a result of defective pollen or ovules, or other aberrations.

Stigma: The part of a pistil or style which receives the pollen for effective fertilization, usually distinguished by bearing minute papillae or having a viscid surface.

Stolon: A trailing/horizontal stem, capable of forming roots and shoots from its node.

Stoma (plural stomata): A minute orifice or mouth like opening between two guard cells in the epidermis, particularly on the lower surface of the leaves, through which gaseous interchange between the atmosphere and the intercellular spaces of the parenchyma is affected.

Strain: A group of individuals of common origin, generally denoting a more narrowly defined group than a cultivar, a group of similar individuals within a variety.

Strain building: Improvement of cross fertilizing plants by any one of a number of methods of selection.

Style: The stalk connecting the ovary and the stigma.

Substitution line: A line in which a pair of chromosome has been replaced by a pair from another variety of the same species.

Succulent: Juicy; fleshy.

Sucker: A vegetative shoot of subterranean origin.

Susceptible: Characteristic of a host plant such that it is incapable of suppressing or retarding an injurious pathogen or other factor.

Syn_0, Syn_1, Syn_2: Symbols for designating the original synthetic population, first synthetic generation (progeny of Syn_0), and second synthetic generation (progeny of Syn_1).

Symmetric hybrid: Stable allopolyploid hybrid.

Synapsis: Conjugation at pachytene and zygotene of homologous chromosomes.

Synthetic allopolyploid: Allopolyploid produced experimentally by man.

Synthetic variety: A variety produced by crossing inter se a number of genotypes selected for good combining ability in all possible hybrid combinations, with subsequent maintenance of the variety by open pollination.

T: Abbreviation for thymidine (2, 6-dihydroxy-5methyl-pyrimidine, 5-methyluracil), a pyrimidine base characteristic component of DNA.

Taq polymerase: Thermo-stable DNA polymerase from *Thermus aquaticus*, a thermophylic bacterium found in hot springs and underwater vents, widely used in nucleic acid amplification.

T-DNA: Segment of DNA in Ti plasmid of *Agrobacterium tumefaciens* that is transferred to plant cells and inserted into the chromosomes of the plant.

Test row: Testing material for resistance against a disease/pest.

Three way cross: A cross between a single cross and an inbred.

Tissue culture: The development of an entire organism plant cells and tissues in vitro or artificial media.

Tolerance: ability of a host to avoid the loss in productivity although it has been infected by a pathogen.

Top cross: An outcross of selections, clones, lines or inbreds to a common pollen parent, in corn, commonly an inbred-variety cross.

Transcription: Process through which RNA is formed along a DNA template.

Transgene: A foreign or modified gene that has been introduced into an organism.

Transgenic: A term applied to an organism that has been altered by introducing DNA molecules into them.

Transgressive variation: Appearance in the F_2 (or later) generation of individuals showing more extreme development of a trait than either of the original parents.

Translation: The conversion of messenger RNA sequence into a protein.

Trihybrid: Offspring from homozygous parents differing in three pairs of genes.

Trimonoecious: Plant having hermaphrodite, male and female flowers.

Trioecious: Plant having hermaphrodite, male and female flowers on different individuals.

True-breeding: An organism homozygous for a trait.

Tuber: A swollen underground stem tip that contains stored food material and serves as a source for vegetative propagation.

Ultraviolet (UV) radiation: The portion of the electromagnetic spectrum wave-lengths from about 1 to 350 nm between ionizing radiation and visible light. UV is absorbed by DNA and is highly mutagenic to unicellular organisms and to the epidermal cells of the multi-cellular organisms.

Umbel: A flat topped inflorescence, like sunflower, in which all of the flowers are borne on pedicel of approximately equal length and all arise from the apex of the main axis.

Unconscious selection: Non-intentional human selection, synonymous with automatic selection.

Unsaturated fatty acid: A fatty acid that has a double bond between the carbon atoms at one or more places in the carbon chain; hydrogen can be added at the site of the double bond.

Upland rice: rice grown like wheat or other cereals in non-bounded fields.

UPOV convention: International Convention for Protection of New Varieties of Plants, revised in 1972, 1978 and 1991 has gradually strengthened the rights of plant breeders.

U triangle: It is a triangle sowing relationships between cultivated species of genus *Brassica* as elaborated by N. U. in 1935.

Variegation: Mosaic phenotype is a widespread phenomenon attributed to plastid variation, genetic instability, instability of the phenotypic expression of the genes, somatic crossing over, and somatic instability of chromosomes (breakage-fusion-bridge-cycle), and transposable elements.

Variety: A subdivision of a species. An agricultural variety is a group of similar plants that by structural features and performance can be identified from other varieties within the same species.

Vernalization: The treatment of seeds before sowing to hasten flowering. It may be accomplished in certain species by the exposure of germinating seeds to temperature slightly above freezing.

Vertical disease resistance: Resistance conditioned by one or a few qualitative genes.

Vertical gene transfer: Introduction of genes from one plant species into another through introgressive hybridization.

Viability: Ability to live, grow, and develop.

Virulence: Capacity of a pathogen to overcome the defense of a host.

Virus: Infectious nucleoprotein, nucleic acid contained in a protein coat.

Wide cross: Cross between two species of the same genus of different genera. It is similar to distant cross.

Wild type gene: The allele commonly found in nature or arbitrarily designated as normal.

Working collection: It is short term storage of germplasm, as seed or clones that breeders can regularly tap for use in development of new crop varieties.

World collection: A collection of germplasm of a particular species from different geographic locations globally, used as source material in plant breeding.

Xenia: Effect of pollen on the embryo and endosperm.

Xenogamy: Cross-pollination.

X-ray: Penetrating electromagnetic radiations having wave length shorter than those of UV or visible light.

Yield plateau: A temporary stable state in yield reached in course of increased production.

Z-DNA: The left-handed double helix that forms in GC-rich DNA molecules. The Z refers to the zigzag paths of the sugar-phosphate in this form of DNA.

Zygote: The cell produced by the union of two mature sex cells (gametes) in reproduction; also used in genetics to designate the individual developing from such a cell.

1

Commercial Crop Breeding Concepts and Historical Developments

Commercial Crop Breeding

The term commercial crop breeding is straightaway equivalent to 'commercial plant breeding' and remains largely plant breeding with slight twist that commercial plant breeding aims at delivering commercial products to be delivered to the end users that is the farmers as per the need of the consumers/ markets. Thus, the farmers plant those products (open-pollinated cultivars and hybrids) which are easily accepted by the consumers and which are easy to be handled by the traders as the products reach to the traders and then from traders to the retailers and ultimately to the consumers. Along with consumers' acceptance, these kinds of products must meet certain level of yield, maturity and resistance/tolerance to diseases and pests. Accordingly, these are designed and bred by the commercial seed companies which design and deliver the commercial cultivars as per demand of the markets. Private seed companies operate on a business model and generate their own revenue to keep the companies running and therefore these seed companies are bound to design and breed commercially viable products only and can ill afford to chase academic butterflies, although these companies encourage innovations for which certain portion of the R & D budget is earmarked. Thus, all plant breeding activities which occur at the research farm and laboratory of private seed companies aimed at designing and delivering commercially viable field crop products are called as commercial crop breeding activities.

The term commercial crop/plant breeding is generally not in use in public sector research institutes and the agricultural universities and certainly not in class room teachings and discussions while teaching courses on plant breeding and related courses under the domain of plant breeding. There the focus is on theory in contrast to applications which form the backbone of commercial crop breeding as practiced by the commercial organizations such as private sector seed companies. Therefore, a working definition of commercial crop or more broadly plant breeding could be as follows.

"The art and science of manipulating the heredity of plants for specific purpose of designing and creating a commercial product (varieties/hybrids/etc) as per need of consumers and traders and accordingly as adopted by commercial seed organizations/companies in private sector".

Commercial plant breeders recognize the role of market forces while choosing the crops, types of the products (open-pollinated seed/hybrid seed/etc) to be developed and delivered in least possible time and accordingly refine and innovate plant breeding and testing procedures to meet the challenges posed by the consumers and naturally the markets. It follows thus that commercial plant breeding is largely true plant breeding as ever defined and understood but with obvious change in goals in terms of products to be developed and commercialized and strategies to be adopted to deliver a target product as per predefined time scale. Therefore, definition and understanding of the term plant breeding cannot be ignored if one has to fully imbibe the meaning of the term commercial crop breeding.

Defining Plant Breeding

The simplest definition of breeding given by Allard (1960) is that breeding is the art and science of changing plants or animals genetically. This definition can be changed depending upon whether it is being used in the context of plants or animals. Thus, plant breeding can be defined as the art and science of changing plants genetically. In most generalized and broad terms, plant breeding is the genetic adjustment of plants to the service of man. Several other definitions of plant breeding have been put forward, such as 'the art and science of improving the heredity of plants for the benefit of humankind' (J. M. Poehlman), or 'evolution directed by the will of man' (N. I. Vavilov). Bernardo (2002), however, offers the most universal definition: 'Plant breeding is the science, art, and business of improving plants for human benefit.' This definition takes into consideration the requirements of commercial crop breeding.

Lately, crop improvement has been used quite often as replacement of the term plant breeding and therefore plant breeding has also been defined as the improvement of crop plants as the result of process of evolution directed by man for his own ends. This definition of bringing evolution in the picture is based on the fact that the basic principles of plant breeding are rooted in the principles of evolution. Definition of plant breeding will remain incomplete unless definition by Smith (1966) is reproduced. Smith (1966) defined plant breeding as the art and science of improving the genetic pattern of plants in relation to their economic use. Although selection is not included in Smith's definition of plant breeding, it is a primary activity in all plant breeding programmes. Selection requires making a choice, and in a breeding programme there are many choices to be made, such as choice of parental germplasm, choice of breeding methods, choice of genotypes for testing, choice of testing procedures, and choice of particular cultivar released for commercial use (Hallauer, 1981).

Plant breeding is a deliberate effort by human to nudge nature, with respect to the heredity of plants, to an advantage. The changes made in plants are permanent and heritable and the professionals who carry out this task are called plant breeders. This effort at adjusting the status quo is instigated by a desire of humans to improve certain aspects of plants to perform new roles or enhance existing ones. As a result, the term plant breeding is often used synonymously with plant improvement in recent years. Normally the term plant breeding connotes the involvement of sexual process in effecting a desired change; however, it is also true that modern plant breeding also includes the manipulation of asexually reproducing plants (plants that do not reproduce through the sexual process). Plant breeding is therefore about manipulating plant attributes, structure, and composition, to make them more useful to humans.

Gepts and Hancock (2006) have defined plant breeding as an applied, multidisciplinary science. It is the application of genetic principles and practices associated with the development of cultivars more suited to the needs of human than the ability to survive in the wild; it uses knowledge from agronomy, botany, genetics, physiology, pathology, entomology, biochemistry and statistics. Of particular importance is the ability to transfer, in addition to major genes, large suites of genes conditioning quantitative traits such as productivity and other traits of interest to humans. The ultimate outcome of plant breeding is mainly improved cultivars. Therefore, plant breeding is primarily an organismal science even though it is eminently suited to translate information at molecular level (DNA sequences, protein products) into economically important phenotypes. The traditional definition of plant breeding includes only those scientists who develop new cultivars and improved germplasm, however, many feel this definition should be expanded to include scientists who contribute to crop improvement through breeding research. This concept has now become a reality as the specific crop improvement research groups include scientists from other related disciplines although the plant breeder remains in the lead role. The art element of the definition of plant breeding has now faded away in view of new breeding tools based on precision and perfect genotyping and phenotyping.

The simplest operational definition of plant breeding has been given by Burton (1981) and that says that a successful breeding programme consists of a series of activities summarized in six words: variate, isolate, evaluate, intermate, multiply and disseminate.

Historical Developments in Crop Improvement in India Related to Commercial Crop Breeding

An exhaustive review entitle "Crop Science" authored by the current luminaries of plant breeding, plant genetics and molecular biology, namely, Prof. R. B. Singh, Prof. (Mrs.) Renu Chopra Khanna, Prof. Anirudh K. Singh, Prof. S. Gopala Krishnan, Prof. Nagendra K. Singh, Prof. K. V. Prabhu, Prof. Ashok K. Singh, Prof. K. C. Bansal and Prof. M. Mahadevappa has appeared in a publication entitled "100 Years of Agricultural Sciences in India" in 2015 edited by Prof. R. B. Singh and published by National Academy of Agricultural Sciences, New Delhi. Chapter 1 "Crop Science" covers Genetics and Plant Breeding, Plant Genetic Resources, Plant Biotechnology, Plant Physiology, Plant Biochemistry and Seed Technology. However, the present write-up is limited to the outstanding historical events related to crop breeding and includes few outstanding achievements of conventional plant breeding and molecular breeding to serve as an eye opening account of outstanding past crop breeding achievements including recently used biotech approaches and these are certainly of great relevance to modern day commercial plant breeders (Singh, 2015). For this chapter, only 3 aspects from originally listed 8 aspects in the review under reference have been taken. Further, crops relevant to seed industry only have been considered.

1. Genetic improvement of crops through conventional plant breeding
2. Plant biotechnology, molecular marker-aided selection (MAS), genomics, molecular breeding and other modern "non-conventional" approaches for crop improvement
3. The flow of quality seed from breeders' plots to farmers' fields

Genetic Improvement of Crops Through Conventional Plant Breeding

Systematic scientific efforts to improve genetic potential of crops in India were initiated in 1905 at the Imperial/Indian Agricultural Research Institute. In 1929, with the establishment of the Indian Council of Agricultural Research (ICAR), several associated developments took place in the first half of the 20th century leading to establishment of several crop-based institutes (sugarcane, cotton, jute, rice, oilseeds, tobacco, potato) and multidisciplinary institutes such as the Indian Agricultural Research Institute (IARI), Central Arid Zone Research Institute (CAZRI), and Central Plantation Crops Research Institute (CPCRI). Additionally, a Project on Intensification of Regional Research in Cotton, Oilseeds and Millets (PIRRCOM) was implemented during 1950s with establishment of several research stations in different regions.

This was followed by major developments in 1960-70 with the reorganization of ICAR set-up, ensuing establishment and strengthening of agricultural research institutes, establishment of a network of All-India Coordinated Crop Improvement Projects, growth of State Agricultural Universities, establishment of National Research Centres (NRCs) and Project Directorates (Randhawa, 1979). Further, during this period, the International Crop Research Institute for the Semi-Arid Tropics (ICRISAT), International Rice Research Institute (IRRI), International Centre for Maize and Wheat Research (Centro Internacional de Mejoramiento deMaiz y Trigo, CIMMYT) were also established. The last more than 100 years of science-based genetic improvement of crops can be broadly classified into three phases - (i) the pre-Green Revolution era (1905-1965), which saw genetic improvement with the application of basic breeding principles of introduction, adaptation and selection with initiation of application of Mendelian Laws; (ii) the Green Revolution era (1965-1985), which saw introduction of photo-insensitive, semi-dwarf, high-yielding wheat and rice genotypes/varieties responsive to high input, and establishment of a comprehensive seed system, and (iii) the post-Green Revolution era (1985 onwards) that has strengthened earlier gains with more focus on resistance breeding, successful development and use of hybrid technology in more and more crops, and application of marker assisted selection in commercial crop breeding.

Pre-Green Revolution Era (1905 to 1965)

This was the initial period of genetic improvement of crop species with the application of basic principles of genetics and plant breeding. During this period, major concentration was on capacity-building, including physical, technological and biological (germplasm), and genetic improvement of food-crops to increase productivity, particularly of the staple food crops, such as cereals, to achieve food security. The major plant breeding procedures utilized during this era were introduction, selection, hybridization followed by selection in the segregating generations, hybrid breeding especially in maize and mutation breeding, particularly in pulses and oilseeds, primarily aiming at improving productivity. The targeted crops were mainly wheat, rice, maize; sorghum and pearl millet for improvement.

Wheat

As mentioned earlier, systematic genetic improvement by breeding in wheat was initiated by Sir Albert Howard and Mrs. Howard at the then Imperial Agricultural Research Institute, Pusa, Bihar, leading to development of several varieties. Howards developed varieties with better yield and quality by selection from local mixed types and two wheat varieties, namely, New Pusa 4 (NP 4) and NP 12 became very popular. Introductions did not play an important role in early breeding, however, one introduction from Australia, "Ridley", fairly resistant to rust; stiff strawed and adapted to medium elevation became popular and was extensively grown. Introductions were often used as source of resistance in breeding programme. In 1907, wheat- breeding research that had already been initiated in 1905 at Pusa, was also initiated at Quetta, Layallpur, Punjab (now in Pakistan). Sir Howard grew two crops of wheat in a year, between November and April at Pusa and during June to September at Quetta, thereby reducing breeding period to half for a variety. Reducing breeding cycle continues even today.

In 1940s, B. P. Pal also emulated this by alternately growing two crops of wheat-breeding materials in a year at Delhi and at Wellington in Nilgiri Hills of Tamil Nadu, and in 1950s, at Delhi and Lahaul Spiti. Similar system of growing two crops of wheat became popular in CIMMYT, Mexico in 1950s when Nobel Laureate Norman Borlaug started raising two wheat crops between El-batan and Obregon in Mexico during the same year, which was known internationally as 'shuttle breeding'. The term 'shuttle breeding' was coined by Borlaug with its implication in breeding for wider adaptability although others had also used this. Based on the Mendelian genetics and pure-line theory, inter-varietal hybridization in bread- wheat in 1920s resulted in the development of world famous high-grain quality wheat varieties, C 518 and C 591, by Choudhary Ram Dhan Singh of Punjab. Among some of the other varieties developed by hybridization was NP 52, which originated from the cross between Pusa 6 and Punjab 9, and became very popular in North India. K. C. Mehta and B. P. Pal initiated in 1928 wheat rust resistance breeding using diverse genes from Khapli.

Based on the principles of mutation breeding, an awned mutant, NP 836, was selected from X-ray radiated variety NP 799 at the IARI. The awned character was monogenic, and the mutant resembled parent line in morphological characters, grain shape, quality and rust resistance (Pal and Swaminathan, 1960). Upadhya and Swaminathan (1963) reported a number of gene transfers from wild relatives into cultivated wheat, without any aberration or deleterious behaviour.

Rice

Initial rice breeding was mostly devoted to improvement in locally adapted large numbers of landraces/cultivars using both mass and pure- line selections. Among these, GEB 24 (popularly known as Kichli Sambha) was one of the landmark rice varieties, which was a spontaneous mutant from a traditional rice variety Konamani, released in 1921 in Tamil Nadu. It proved to be a useful variety and was extensively used in the development of several improved varieties such as TKM 6, which in turn has been involved as a parent in the development of internationally popular varieties, IR 20, IR 26 and IR 36.

Hybridization as a method of breeding in rice was initiated in the early part of 20th century. To improve varietal response to chemical fertilization, a coordinated *indica* x *japonica* hybridization project was initiated by the Food and Agriculture Organization (FAO) at the Central Rice Research Institute (CRRI), Cuttack, in 1950 to combine high nitrogen response and yield potential with insect and pest resistance into *indica* background (Parthasarthy, 1954), resulting in release of varieties such as 'ADT 27' in Tamil Nadu and 'Mahsuri' in Malaysia. Breeding for resistance to blast in rice was initiated in India in late 1920s by K. Ramaiah and K. Ramaswamy in 1936 that bred for resistance to *Pyricularia oryzae*, while varieties like BAM 10, T 141 and CH 13 were reported to be resistant to brown spot.

Mutation breeding was also initiated with first report of induced mutation by K. Ramaiah and N. Parthasarathy in 1936. The earliest known case of irradiation induced mutation to produce dwarf genotypes was achieved in India, where in seeds of GEB 24 were exposed to X-rays of different intensities during 1932 by K. Ramaiah and N. Parthasarathy. An economic type of mutant was selected, which was shorter with better tillering than GEB 24 and was adapted to fertile soils. Unfortunately, fertilizers were not commonly used at that time, and the critical tests of fertilizer response were not conducted until later. Further, this variety did not become popular with farmers as they preferred longer straw for animal-feed. A mutant was selected following X-ray treatment of T 141 at the Orissa University of Agriculture and Technology; the mutant strain had shorter and stiffer straw than the parent. Varieties like Ishwarkora and TKM 6 were found resistant to stem-borer because of their stem thickness and hardiness.

Attempting to use wild relatives in rice improvement, based on the crosses between *Oryza sativa* f. *spontanea* x *O. sativa*, Indian scientists were the first to report role of the cytoplasm in causing male sterility in rice and potential use of the system in breeding hybrid varieties (Sampath and Mohanty, 1954). Richharia (1962) advocated clonal propagation as the practical means for exploiting hybrid vigour in rice.

Maize

In maize, in earlier times, mass selection was the principal method of breeding. It was used both for maintaining existing varieties and for developing new varieties with specific characters. Hybrid maize, involving inbreeding to establish pure-lines, and hybridizing pure-lines to produce uniform hybrids with favourable gene combinations resulted in inherent high yield. In India, inbred lines were developed and released through the Indian Coordinated Maize Improvement Scheme (ICMIS). They were prefixed with CM letters (indicating Coordinated Maize) — CM 109, CM 202, etc. In maize, first commercial hybrid (Punjab Hybrid No. 1) was developed in 1956 in Punjab. Later, a large number of inbred lines were developed both by public and private institutions and progress was made with single crosses, double crosses and other crosses, leading to development and release of two three-way cross hybrids, Ganga Safed Hybrid Makka 2 and Hi-Starch Hybrid Makka. Both hybrids involved a single cross, which was pollinated by an open-cross variety. For eating purpose, flint varieties were preferred to dent types. For this reason, except high-starch types, most hybrids released by the ICMIS were orange-yellow flint types. However, the development of starch industry in the country led to increased use and distribution of two white dent-type hybrids for starch yield as reported by Paliwal (1964). Sweet Maize Hybrid No. 1 was produced in India by combining inbred lines selected from crosses between US sweet types and Indian flint types.

Pearl Millet

In pearl millet, initially most of the Indian varieties were developed by selection from local types. Many were developed by a single plant selection in different areas of cultivation. In Madras state, synthetics were developed by mixing seed of six inbred lines in equal proportion and by mixing seeds of 15 possible single crosses between 6 inbred lines. For exploitation of heterosis, before cytoplasmic male-sterile lines were discovered, attempts were made for utilizing hybrid vigour in Madras and Bombay states. Several sources of cytoplasmic male sterility were reported in the meantime (Menon, 1959).

Sorghum

In sorghum, an often cross-pollinated crop, in earlier days, methods that were followed for genetic improvement were similar to those used for self-pollinated crops, like introduction, selection and hybridization. The selections made were generally poor in productivity. Hybridization was taken up for developing dual-purpose varieties— combining good seed quality with fodder potential. Co 20, a striga-resistant variety was produced in Madras by transferring its resistance from an African line to a local type (Shanmugasundaram and Venkatraman, 1964). Backcrossing was also used in breeding some varieties for acquisition of specific desirable traits, like yellow endosperm to parent line. At this stage, breeding for resistance to diseases was also launched.

Grain Legumes

Most pulses are self-pollinated crops; therefore, the breeding methods used in the past were as applicable to self-pollinated crops, namely, introduction, selection and hybridization, followed by pedigree selection. In addition, special techniques like mutation through irradiation were also attempted. In the initial stages, pure-line selection from indigenous varieties/local types/landraces was the principal method of pulse breeding. India is one of the centres of origin/ diversity of the major pulses such as chickpea, and this provided opportunity to select from local diverse types. Chickpea variety G 24, a pure-line selection from the local type, was released in Punjab in 1958, with higher yield, drought resistance, early maturity and wilt resistance. Other varieties developed through selection from local landraces during 1947 - 1960 were RS 10, Annegeri, etc. Similarly in pigeonpea, a drought- resistant strain, SA 1, was developed in Madras through pure-line selection.

The first systematic crop improvement effort in lentil was made in 1924, when 66 pure-lines designated as 'Type 1' to 'Type 66' were selected at the erstwhile Imperial Agricultural Research Institute, Pusa, Bihar. Introduction was another method, and introduced varieties were either used for direct cultivation/ adaptation or were used in hybridization programme.

Shining Mung No.1, a green gram variety, was developed by selection in Punjab from an introduced Chinese variety (Mehta and Sahai, 1955). A bold-seeded, white-grained African Chickpea variety 'Rabat' was crossed in Punjab with a local variety Pb. 7 to develop white, bold-seeded variety C 104 (Athwal and Bajwa, 1965). Later, hybridization was used to develop improved varieties by combining good characteristics. The chickpea variety C 1234 was developed through hybridization between Pb 7 and an exotic variety F 8. Genetics of single- and double-seeded pods was studied, which indicated that single-seeded pod showed dominance over double-seeded pod. Attempts were made to improve yield by transferring double-seeded pod trait.

Mutations were induced in *Cicer*, *Cajanus* and *Phaseolus* by irradiation (Jana, 1962; Athwal, 1963). Though there was no research on adaptability, in Punjab variety S 26 was recommended for rainfed areas, while S 33 was developed for irrigated or areas with adequate rainfall. Black gram variety Kulu Mash 4 was recommended for hilly and sub-mountainous tracts. With regard to resistance breeding, pigeonpea variety S 1 was reported drought resistant in Madras. Resistance to wilt, caused by *Fusarium udum*, was reported in N P 41, N P 51 and N P 80 (Singh, 1957) and a cross between N P 51 and N P 24 gave rise to the development of four highly wilt - resistant selections, N P (WR) 15, N P (WR) 16, N P (WR) 19 and N P (WR) 38 as reported by Deshpande et al., (1963) and Randhawa (1963).

The wild relatives of *Cajanus*, *C. lineata* and *C. sericea* (then called *Atylosia*) were also-reported resistant to wilt and were used in resistance breeding (Singh, 1957). Wilt resistance was found genetically controlled by a pair of duplicate dominant genes and also by multiple genes (see review by Singh et al., 2015). The first report on male-sterility in pigeonpea was published by Deshmukh (1959).

Quality breeding was initiated in chickpea, where bold-seeded, white 'Kabuli' types are preferred over brown small-seeded Desi type, and variety C 104 was developed in Punjab from a cross between Pb 7 and Rabat. Pb 7 is a Desi type, while Rabat is an African Kabuli type. The evolved variety C104 is white, bold -seeded and yields equal to Pb 7.

Jain (1975) proposed the concept of plant type in pulses, emphasizing the breeding of high- yielding pulse varieties for a higher level of management, based on the genetic reconstruction of the existing plant types, specifically with higher harvest index, response to increased plant population per unit area and early maturity.

Rapeseed and Mustard

In rapeseed- mustard, there are species with self-fertile forms and also with self-sterile forms. Even in self-fertile forms, cross-pollination ranges from 5 to 15%. Therefore, these crops are categorized as often cross-pollinated crops like cotton, and the breeding procedures used include mass selection, progeny selection and hybridization for combining useful genes. In self-sterile forms, breeding procedures used were similar to cross-pollinated crops, including mass selection, recurrent selection, development of synthetics and hybrids. Initially, genetic improvement involved purification of landraces.

Scientific breeding of rapeseed -mustard started at Layallpur of undivided Punjab, releasing RL 18 (Raya Layallpur 18) in 1937 and L1 of yellow sarson (mustard) through selection. Development of synthetics using poly cross technique to identify lines with superior combining ability was found to be a practical and effective method (Maini and Ghai, 1964; Sikka and Rajan, 1957). Polyploidy in 'toria' was tried at the IARI by N. Parthasarthy, S. S. Rajan and

Y. R. Ahuja but could not become a commercial success due to poor seed set. Between 1947 and 67, a number of varieties of mustard (Laha 101, Varuna, Durgamati, Patan mustard), of toria (Abohar, BR 23, M 27, T 9, ITSA, T36, DK 1), of brown sarson (BSA, BSG, BSH 1, BS 2, BS 65, BS 70), of yellow sarson (T 151, Patan sarson, YSPb 24, T42) and of Taramira (ITSA) were developed (Chauhan et al., 2010).

Groundnut

Groundnut (*Arachis hypogaea*) was introduced into India through eastern coast from countries of Southeast Asia, originating in Brazil followed by later introductions from Africa and the USA. Most stations and the IARI involved in groundnut breeding maintained exotic collections (Chandrasekharan et al., 1960). Introduction and selection still remained main methods of genetic improvement. Selections were made from both exotic and local strains. Pure-line selection from local types led to development of RS1, while selection from exotic strains from Brazil led to development of RSB 87 in Rajasthan (Bhatnagar et al., 1964). Radiation was also used for induction of variations during 1950s. Varieties resistant to diseases, 5203 of Gwalior; G 0120, G 1032 and G 0607 of Mysore and exotic 4 of Indore were identified.

Soybean

In soybean, introduction and adaptation of varieties from the USA was the major activity during the period. Varieties from southern USA were better performers in India because of climatic/photoperiod similarity.

Cotton

Cotton is the most important natural source for textile industry. Archeological remains suggest production and spinning of cotton in India since long, supporting India as the place of origin and domestication of diploid Asiatic or desi cotton. Originally, only indigenous desi cotton, belonging to two indigenous species, *Gossypium arboretum* and *G. herbaceum* with short staple length, were cultivated. The long staple *G. hirsutum* or American upland cotton was introduced into India in 1790 via Bombay, and introduction continued throughout the 19th and early 20th century. Upland type with more leaf hairs persisted in fields of Desi cotton, which formed the basis of selection and successful establishment of American cotton as a crop; for example Punjab-American Upland cotton in north India (Hutchinson,1962).

Cambodia cotton, which had long dense hairs, was able to survive jassids attacks. It was a vigorous strain, which yielded well and spread into south India after its original introduction in 1906. Selection was practised in cotton both for maintenance of varieties and development of new varieties. Most early work was based on a single plant selection. The Parbhani-American 1 was selected in Bombay in 1932 as a result of the single plant selection from farmer's field. Similarly, Cambodia 1 and Cambodia 2 were selected in Madras from Cambodia cotton introductions (Ramachandran, 1965). Mass selection was practised by bulking seed from open-pollinated or selfed plants, based on appearance. In Punjab, mass selection was used for purification, improvement and maintenance of upland cotton variety 289 F/K 25 (Kohli, 1952). During pre-independence era, a stress-tolerant variety, Labh Singh Selection (LSS), selected from local collections was popular. Similar selections, 216 F and H 14 *hirsutum* were ruling varieties in 1950s and 1960s in the north-west zone. J 34, a variety developed through hybridization and mass-pedigree selection at the Punjab Agricultural University, Jalandhar, Cotton Research Station, was a reference variety since late 1960s for price fixation of medium-staple American cottons.

During 1960s in Gujarat and in adjoining provinces, G 67 was a leading variety with very high ginning out turn (GOT) of about 50%. Later, many varieties were developed by hybridization — Virnar, H 420, Digvijay, Co 4 and several others. Co 4 was developed from a cross of Co 2, selected from original *hirsutum* Cambodia bulk introduced in 1906, and A-12, a *hirsutum* type, selected from an introduction from Uganda. MCU 1 was developed as a reselection from Co 4. Crosses between Virnar, H 420 and Co 4 were used to combine desirable characteristics into a single strain. Interspecific crosses were used to combine desirable genes from two or more species, both at diploid and tetraploid level. The diploid species, *G. anomalum*, was used to improve fibre fineness and strength of *G. arboreum* (Santhanam and Krishnamourty, 1964).

At tetraploid level, crosses between *G. hirsutum* and *G. barbedense* in Madras resulted in the development of variety MCU 2, by crossing MCU 1, a *G. hirsutum* variety, with a long staple *G. barbedense*. Some success was achieved between diploid and tetraploid species crosses. A long staple variety Ak 8401 was developed from a cross between *G. arboretum* and *G. anomalum*. Variety 170-Co. 2, a cross between *G. hirsutum* and *G. arboretum*, and 134-Co.2 M, a cross between *G. hirsutum* and *G. herbaceum*, were released in Gujarat (Sikka and Joshi, 1960a).

Backcrossing was used for incorporating specific genes and Vijay, Kalyan and Digvijay varieties were developed by backcrossing in the year 1943, 1947 and 1956, respectively (Sikka and Joshi, 1960a). Some important genetic resources, commonly used in basic and applied cotton research, were Super Okra, American Nectar-less 145, Red AK, Acala Glandless, and AET 5.

Similarly mutation breeding, though on a limited scale, was attempted and some natural and induced mutants were reported. A cotton-plant with increased leaf hair density (40-50%) was obtained after X-ray irradiation of Mescilla Acala variety (Jagatheesan and Shastry, 1963). Cotton Indore 2 was developed from X-ray induced mutation of Malwa Upland 4 in Madhya Pradesh, and was released for cultivation in 1950 (Sikka and Joshi, 1960a).

From yield point of view, production of varieties with high lint fibre was the main objective in genetic improvement. Among the Indian varieties, Rosea 231 and Ganganagar in case of *arboreum* and Digvijay in *herbaceum* had high lint percentage, while 134-Co2-M in *Hirsutum* had large boll size. In resistance breeding, there were attempts for integration of resistance against Fusarium wilt. The resistant varieties included H 420 and Virnar in *G. arboreum*; K. F. T., Digvijay and Jayawant in *G. herbaceum* and Co 2 in *G. hirsutum* as reported by Santhanam et al., (1964). For bacterial wilt, resistant varieties developed were from crosses between *G. hirsutum* and *G. herbaceum* (Santhanam et al., 1964; Sikka and Joshi, 1960a). Co. 2 in Madras and Laxmi in Mysore were reported to be resistant varieties to red leaf blight (Sikka and Joshi, 1960a).

Most of the varieties with resistance to jassids were developed by incorporating increased hair density which created a mechanical barrier against insect infestation. Substantial work was done on the genetics of hairiness; identifying 6 genes responsible for its control. A major gene, H1, was identified in a few Indian varieties, like Malwa Upland 8b, Cambodia UA7 29, belonging to *G. hirsutum*, and Wagad 8 of *G. herbaceum*. Yield improvement was also achieved through interspecific hybridization and backcrossing. In upland cotton, varieties like Arogya, PKV 081, Rajat, Gujarat 7, MCU 2, MCU 5, Deviraj, Devitej, Khandwa 1, Khandwa 2 and Badnawar were developed. Similarly, several commercial hybrids were developed in tetraploid and diploid cultivated species. Important tetraploid hybrids (*G. hirsutum* x *G. barbadense*) included Varalaxmi, released for the Central Cotton- growing zone. Four hybrids — GDH 7, GDH 9 (cultivated in Gujarat), DDH 2 (cultivated in Karnataka) and Pha 46 (cultivated in Marathwada region of Maharashtra), were from crosses between *G. herbaceum* and *G. arboreum*. These cultivars are still being used as parents in cotton breeding.

The Green Revolution Era (1965 to 1985)

Semi-dwarf Wheat and Rice High Yielding Varieties (HYVs)

Spearheaded by M. S. Swaminathan in 1960s, development of semi-dwarf, high yielding, input-responsive, photo-thermo-insensitive, and high harvest index wheat and rice varieties for Indian conditions triggered unprecedented yield increase which was dubbed as Green Revolution in India (Swaminathan, 1968, 2006; Shastry, 1972). The Indian wheat and rice breeders most judiciously deployed and harnessed internationally exchanged dwarfing genes, and, in four years, (1964 -1968), added genetic gains of a magnitude similar to that attained during the foregone four-thousand years. This was attributed to the "symphony" of science, technology, policies, political will and farmers' enthusiasm. The term Green Revolution is so commonly used by all (students, scientists, farmers and policy makers) that it needs a bit more elaboration to put the things in the right perspective.

Green Revolution signifies quantum jump in production and productivity of wheat and rice between 1965 and 1975 in developing countries mainly in middle-east and south-east Asia on account of application of modern principles of plant breeding and genetics in an aggressive manner. Perhaps the most notable essay on food and population dynamics was written by Thomas Malthus in 1798 entitled "Essay on the Principles of Population" where he identified the geometric role of natural population increase in outrunning subsistence food supplies. He observed that unchecked by environmental and social constraints, it appears that human populations double every 25 years, regardless of the initial population size. According to this proposition, population increase was supposed to be geometric whereas food supply at the best was arithmetic. Obviously, this proposition was full of pessimism about feeding ever growing populations. Fortunately, mitigating factors such as technological advances, advances in agricultural production, changes in socioeconomics, and political thinking of modern society, has enable this dire prophesy to remain unfulfilled (Acquaah, 2007).

However, the advantages of technological advances in 20th century primarily could reach to the industrial countries leaving most of the developing countries still struggling with hunger and malnutrition. In 1967, a report by the US President's Science Advisory Committee came to the grim conclusion that "the scale, severity, and duration of the world food problem are so great that a massive, long range, innovative effort unprecedented in human history will be required to master it". The Rockefeller and Ford Foundation, in response to this challenge, proceeded to establish the first international agricultural system to help transfer the agricultural technologies of the developed countries to the developing countries. These humble beginnings led to a dramatic impact on food production in the third word, especially Asia which could be dubbed the Green Revolution. The term 'Green Revolution' was coined by William S. Goud, former Administrator of the United States Agency for International Development (USAID, in a speech to the Society of International Development in 1968, to signify the phenomenal gains in food grain production achieved in 1966 to 1968 especially in India and Pakistan through the introduction of semi-dwarf high yielding wheat varieties from CIMMYT, Mexico.

The Government of India issued a postal stamp in 1968 to highlight wheat revolution. As early as 1968, Swaminathan emphasized the need for pursuing enhancement of productivity in perpetuity without ecological harm —"ever-green revolution"— the term he had coined (Swaminathan, 2007). Genes responsible for photo - and thermo –insensitivity coupled with earliness helped crops adapt to varying photoperiod and temperature regimes from semi-temperate to tropical and sub-tropical conditions, which enabled India to achieve one of the highest cropping intensities in the world.

Identification of genes responsible for absorbing and utilizing nutrients and increasing harvest index from less than 0.2 in early 1900s to more than 0.50 by the mid-1970s was crucial in achieving high economic yields and intensification (Jain and Kulshreshtha, 1976). Through the Green Revolution, India, with an annual food grains production of about 75 million tons in 1966, reached an all-time high production of over 260 million tons in 2014, transforming the country from a ship-to-mouth situation to a self-reliant and self-sufficient food secure nation, enabling it to strive to enact a Right to Food Bill in 2012 – indeed a historic transition (Swaminathan, 2012). The food production has now reached to 285 million tons in 2017-18 signalling still greater impact of Green Revolution.

Initial efforts in wheat resulted in the release of four lines as varieties in 1967 with names — Kalyan Sona, Safed Lerma, Sonalika and Chhoti Lerma. At the same time, Sonora 64 and Lerma Rojo 64A, possessing un-preferred red grains, were subjected to mutagenesis, producing Sharbati Sonora and Pusa Lerma with amber colour grain to meet consumers' acceptance. V.S. Mathur played a key role in the development of a range of very high-yielding and high quality wheat varieties at IARI, New Delhi (Swaminathan, 2013). Contemporary wheat breeders who played crucial role in developing superior varieties of wheat during this period at state agricultural universities were J. P. Srivastava at GBPUAT, Pantnagar, Khem Singh Gill at PAU, Ludhiana, S. M. Gandhi at Research Station, Kota, R. B. Singh at BHU, and several others.

Introduction of 'miracle' rice variety 'IR 8' (tall Peta x Dee-Geo-Woo-Gen) in 1966, developed by the International Rice Research Institute (IRRI), and release of the first Indian dwarf rice variety, 'Jaya' with good yield and high per day productivity in 1968, marked the beginning of the miracle rices that brought rice revolution. In 1966 a cultivar IR 8-283-3 (named IR 8) from IRRI made the flash point in the hybridization programme of *indica* x *japonica* cross. IR 8, an abrupt change from the traditional to the modern plant type is characterized by eight features ((Stoskopf, 1885; Ram, 2019).

1. Semi-dwarf plants ranging from 90-120 cm in height.
2. Short, thick, sturdy, stems that provide lodging resistance even under high nitrogen levels.
3. A high photosynthetic rate related to improved sunlight interception by short, upright leaves of medium width.
4. High tillering capacity, giving rise to more rice panicles per unit area of land.
5. Increase in harvest index from 30 to 50 percent
6. Insensitivity to photoperiod enabling reproductive development regardless of the date of sowing and thus allowing for more than one crop per year.
7. Increased grain yield under high levels of nitrogen fertility.
8. In 1966 dry season at IRRI, IR 8 yielded 9.4 tons/ha and yielded 10.3 tons/ha in replicated experiment at IRRI in the same year, the highest yield recorded for any variety in the replicated experiment in the tropics at that time.

One of the major landmarks in rice improvement during this period in India came with the development of rice variety 'Swarna' (MTU 7029) in 1979 at the Agricultural Research Station (now renamed as Andhra Pradesh Rice Research Institute), Maruteru, through pedigree selection from Vasista × Mahsuri. This mega variety is a standing testimony of rice improvement through breeding; occupying more than 5 million ha of rice cultivation in 1980s and 1990s in India, and was also widely grown in other countries across the world in Bangladesh, Sri Lanka, Nepal, Uganda, Ethiopia and Kenya. Rice variety Co 4 (a selection from Anaikomban) was identified as blast resistant in Tamil Nadu (Parthasarathy, 1972). The science-led gains of the Green Revolution (GR) were protected, consolidated and further enhanced with identification, characterization and mobilization of genes conferring tolerance/resistance against various biotic and abiotic stresses, thereby protecting and fortifying gains of increased productivity/harvest (Nagarajan et al., 1978; Reddy and Rao, 1979).

In 1964 Sonora 64 and Lerma Rojo 64A were released and 18,000 tons of seed imported from CIMMYT's host country, Mexico, through joint efforts of Prof. Swaminathan and Nobel Laureate Norman Borlaug, with the support of the then Prime Minister Smt. Indira Gandhi, Agriculture Minister, Bharat Ratna C. Subramaniam and Agriculture Secretary, Shri Sivaraman. In 1965 S 227 was released as Kalyan Sona, (owned as Kalyan by the IARI and Sona by the PAU), in 1967 Sonalika was released for late-sown irrigated and timely rain-fed conditions as the first widely adapted variety. It was tested as RR 21 at Kanpur, as S 308 by the GBPUAT, Pantnagar, and PAU, and as HD 1553 by the IARI. In 1980 HD 2009, and Arjun, (the first native dwarf variety, well received by farmers of the north-western plains) were released.

Landmark rice varieties of this era are given in Table 1.1 (Singh, 2015).

Table 1.1: Landmark green revolution rice varieties

Year	Technology/Variety
1967	IR 8: The first semi-dwarf, photoperiod insensitive, fertilizer responsive, high-yielding variety, with an average yield of 6.0 tons/ha, maturing in 135 days, was developed at the IRRI from the cross Peta x Dee-Geo-Woo-Gen, and was introduced into India, and had set the tone for "Green Revolution" in rice
1968	Jaya: A semi-tall, photoperiod insensitive variety, yielding 4.0-5.0 tons/ha, bred at the DRR, Hyderabad, from the cross TN1 x T141
1969	Jagannath: The first rice variety developed through mutation breeding at the OUAT, Bhubaneswar, produced an average yield of 5.0 tons/ha
1972	Mahsuri: A product of *indica* x *japonica* hybridization programme, selected in Malaysia; photoperiod insensitive, non-lodging variety, produced 4.5 tons/ha on an average. It was released for Odisha, Andhra Pradesh and Chhattisgarh. It was popularly known as "Ponni" in the southern India. Parents of several mega rice varieties including Swarna (MTU 7029) and BPT 5204 (Sambha Mahsuri)
1978	PR 106: A semi-dwarf, photo-insensitive, high yielding (6.5-8.0 tons/ha) variety, introduced from the IRRI and evaluated in Punjab; established a brand of 'Parmal' rice in export parlance
1978	Nagina 22: It was an upland rice variety tolerant to drought and heat stress with short bold grains and profuse tillering, developed at the Rice Research Station, Nagina, UP. It is used as a donor for drought and heat tolerance worldwide
1979	Swarna (MTU 7029): A widely cultivated high-yielding mega rice variety grown on approximately 5 million ha currently in India, developed from Vasistha x Mahsuri by Shri Rama Chandra Rao at the Regional Rice Research Station, Maruteru, ANGRAU
1980	Sarjoo 52: A semi-dwarf, photo-insensitive variety with good threshability, developed by D. Maurya at NDUAT, Kumarganj, Faizabad (now Ayodhya). It was widely grown in Uttar Pradesh
1981	IR 36: A semi-dwarf, photoperiod insensitive variety developed at the IRRI, Philippines. Resistant to multiple biotic stresses — bacterial blight, gall midge, green leaf hopper, rice tungro virus and grassy stunt virus— and moderately resistant to sheath blight and blast; was dubbed by press as 'Miracle Rice' as it occupied at a given point of time 65% of the global rice area. Developed by the team led by the renowned rice breeder Dr. Gurdev Khush at IRRI, Philippines
1982	CR 1009: A semi-dwarf, photoperiod insensitive variety with an average yield of 5.3 tons/ha, developed at the CRRI, Cuttack

1985 IR 58025A: One of most widely used CMS lines in hybrid rice breeding programme worldwide, developed at the IRRI; utilizing Pusa 167, an elite aromatic long slender grain genotype developed at the IARI, New Delhi

1986 ASD 16: A semi-dwarf, photo-insensitive, early, high-yielding variety, widely adapted in Tamil Nadu

1986 Samba Mahsuri (BPT 5204): A widely cultivated mega rice variety with fine quality medium slender grains. Developed from three-way cross GEB24/TN1//Mahsuri, as student- research programme, guided by Dr. M. V. Reddy, at College of Agriculture, Bapatla, ANGRAU. Its rice is famous even in Silicon Valley, USA

Heterosis Breeding and Hybrid Technology

Exploiting hybrid vigour (heterosis) through non-additive gene effects by producing hybrids between parents with best combining ability for yield as well as for resistance against various diseases and pests, particularly in cross-pollinated crops like maize and pearl millet and often cross-pollinated crop sorghum has complemented dwarfing genes revolutionary technology in wheat and rice ((Paroda, 1995).

Maize

In maize, a 'step ladder' approach was proposed to effectively harness additive and non-additive gene effects (Dhawan, 1965). For economical seed production, double top-crosses (single cross x non-inbred parent) were adopted resulting in release of two double top-cross hybrids. Simultaneously, the concept of composite breeding, being simple was implemented enthusiastically and six early maturing composites were released quickly in 1967. Later, understanding the need for early- maturing composites, Makki Safed 1 (an improved local) was released in 1973, followed by Ageti 76, Tarun and others. So far, nearly 100 composites have been released to meet diverse agro climatic requirements of the country. During 1970s and 80s, large-scale intra-population improvement was taken up to develop better composites (Dhillon et al., 2006). But these composites remained largely confined to limited cultivation areas and failed to make any big impact and almost disappeared.

Sorghum

Among the predominantly self-pollinated species, sorghum is the first cereal staple crop to be commercially exploited for heterosis in the world, and the first success in hybrid sorghum breeding was achieved four decades after the demonstration of heterosis in sorghum by Ganga Prasad Rao of the IARI, who was the first to employ cytoplasmic male-sterility system (CMS) with appropriate fertility restorers in India to release world's first commercial sorghum hybrid, CSH 1. CMS was discovered in 1961. With the adoption of

Combine Kafir 60, a male sterile (MS) inbred, as the common female parent to all immediate crosses and with availability of wide range of germplasm as restorers, hybrid sorghum programme got a fillip (Rao, 1962). Combine Kafir 60 was selected as a male sterile because of its stable sterility, high combining ability, short- stature, lodging resistance and white corneous seed (Rachie, 1965). It was found resistant to stem-borer as well.

Most conventional Indian varieties of sorghum were tall and weak-stemmed, and therefore lodged severely. But hybrids developed with Combine Kafir 60 (ms CK 60A), as female parent, improved lodging resistance. Later, based on the detailed combining ability analysis (Rao and Murty, 1970) and utilizing CMS lines system, a series of CSH (coordinated sorghum hybrids) hybrids were developed, starting with CSH 1 in 1964, to transform kharif sorghum production (Rao et al., 1966; Rao et al., 1986). The adoption of hybrids, based on exotic CMS line (ms CK 60 A), however, suffered from several disadvantages, such as limited heterosis, relatively inferior grain quality and greater susceptibility to shoot-fly, which was overcome by diversification of CMS lines, resulting in development of several improved CMS lines and hybrids. Consequently, production of coarse grains, which are nutritionally rich, was augmented greatly with the exploitation of hybrid vigour by the Indian scientists.

Pearl Millet

The pearl millet hybrid breeding programme followed the same pattern as of sorghum. The world's first commercial hybrid HB 1, released by D. S. Athwal of the PAU, Ludhiana, through the All-India Coordinated Millet Improvement Programme in 1965, used exotic Tift 23A as the source of CMS to be used as female parent from the USA as reported by Athwal (1965). The application of hybrid technology was further strengthened with the isolation of new cytoplasmic male sterile lines at the PAU from different parental materials. Though the male sterile line L 101-A possessed bold seeds and early maturity, it required a different restoration gene than the one associated with restoration of fertility of Tift 23A-based hybrids. Therefore, L 101A could not be exploited. Tift 23A was extensively used because of short stature, profuse tillering, uniform flowering and good combining ability. The possibility of three crops in south India, i.e. Kharif, Rabi and summer, facilitated rapid crossing of inbred lines with male sterile lines, and testing of single -cross hybrids. Large-scale adoption of hybrid varieties evolved by the Indian scientists from 1965 onwards increased pearl millet yields substantially.

Genetic Improvement in Non-Cereals

Grain-legumes

All- India Coordinated Pulse Improvement Project (AICPIP) was established by the ICAR in 1967. It collaborated with the International Crop Research Institute for Semi-Arid Tropics (ICRISAT) for increased access to global germplasm for crop improvement programme. One of the major achievements of the period was development of extra-early pigeon pea varieties, like UPAS 120 from Pantnagar, encouraging wheat planting after harvesting pigeon pea in November. However, this cropping pattern could not become popular. Identification of desirable variants for disease and pest resistance in wild species was strengthened.

In chickpea, through selection RS 11 and through hybridization C 235 and Radhey varieties with high- yield and bold- seed were developed. In other pulses, intensive selections were made from collections obtained from different parts of the country, resulting in the development of many promising varieties in urdbean, mungbean, pea (Type163), etc. In lentil, this period emphasized on the widening of genetic base and adaptability through hybridization programmes, as a result of which 'Pant L 639', the first variety developed by hybridization was released in 1981, followed by the release of many other lentil varieties. Varietal development in lentil got a boost due to availability of large number of germplasm from ICARDA.

Several important varieties were developed through direct selection from local material or landraces in green gram and black gram. In green gram, important varieties developed through selection in different states of India included Type 1, Shining Mung 1, BR 2, etc. In black gram, Type 9 variety was developed through selection from 'Bareilly Local' in 1948 and was extensively used as a parent in the development and release of a large number of varieties (Singh, 1982a, 1982b). Many grain- legume varieties developed through pure- line selection were used in hybridization by plant breeders as a base material for further improvement in disease resistance, yield and yield components. A unique effort, attempting inter-specific hybridization between green gram and black gram, involving former as a female and black gram or derivatives of green gram x black gram as a male, was done. This resulted in the development of 3 varieties (Singh and Ahlawat, 2005).

The first report of genetic male sterility (GMS) in pigeon pea, its inheritance and application in pigeon pea improvement appeared in 1978 (Reddy et al., 1978), and was used for developing first pigeon pea hybrid in India.

Oilseeds

A multidisciplinary research project, All- India Coordinated Research Project on Oilseeds was established in 1967 with major objectives of genetic improvement for oil and seed yield with improved oil quality (low erucic acid) and seed meal (low glucosinolate) and resistance to biotic and abiotic stresses. Pure-line and mass selection resulted in the development of initial varieties, followed by breeding of varieties through hybridization. 'Varuna', a pure-line selection from Banarasi Rai, a local germplasm, released in 1975, was extensively used in the development of as many as 46 varieties of mustard. Rawat and Anand (1979) were the first to report cytoplasmic male sterility in *B. juncea*.

In groundnut, introduction and reselection continued to be the main methods of crop breeding with the release of widely adapted varieties, like JL 24, Kadiri 3, M 13, etc., and initiation of hybridization between selected parents, releasing varieties like J 11. For incorporation of resistance, particularly for leaf- spot, causing maximum yield losses, wild species like *Arachis villosa* were used as source with single factor controlling resistance. Consequently, inter-specific crosses with resistance were successfully produced.

Cotton

In exploitation of hybrid vigour in cotton was suggested in early 1960s. Heterosis was observed in inter-specific crosses for certain features and in varietal crosses of the same species. India is perhaps the only country, which exploited hybrid cotton technology for the first time (Patel, 1971). The important landmarks in cotton improvement include —release of world's first intra-*hirsutum* cotton hybrid 'H 4' in 1970 from Surat, followed by first inter-specific (*G. hirsutum* x *G. barbadense*) cotton hybrid 'Varalaxmi' in 1972 from Dharwad and world's first GMS- based cotton hybrid 'Suguna' in 1978 (Kairon et al., 1998).

Post-Green Revolution Era (1985 to Date)

The genetic gains achieved during the Green Revolution were further consolidated and strengthened with incorporation of genetic resistances for various biotic and abiotic stresses using conventional breeding techniques and modern biotechnological tools such as molecular marker-aided selection and scaling up of hybrid technology with increasing role of private seed companies.

Wheat

In wheat, efforts on gene deployment for saving and fortifying harvest continued with gene pyramiding for multiple resistances to biotic stresses in selected varietal backgrounds, for increased production for food security (Cherukuri et al., 2005; Gupta et al., 2006). In recent years, stem rust race UG 99 has been kept at bay through anticipatory gene deployment (Prashar et al., 2008). One of the most popular bread- wheat varieties, HD 2967 has been bred excluding highly exploited 1B-1R translocation (Veery) for rust resistance by bringing together minor genes for durable leaf rust resistance (Malik et al., 2012). Landmark wheat varieties since 1985 are given in Table 1.2 (Singh, 2015).

Table 1.2: Landmark wheat varieties since 1985

Year	Technology/Variety
1985	HD 2329: First Indian variety that covered 4 million hectares, eventually replaced Kalyan Sona
1985	HD 2285: Replaced late-sown Sonalika
1989	HS 240: High-yielding variety for irrigated and rain-fed areas; occupied >80% of Northern Hills
1995	PBW 343 by the PAU that replaced HD 2329, and occupied 7 million hectares in 2004
1997	Molecular breeding (MAS) in wheat started by the IARI for rust resistance
2005	Indian challenge for breeding resistance against *Ug 99* launched, and India was the first country to overcome *Ug 99* attack of stem-rust using MAS, All HD and HI series varieties released till late are resistant to Ug99
2011	HD 2967: Released and partly replaced PBW 343 with very fast spread covering 8 million ha in its third year of release i.e. in 2014-15
2014	HD 3086: Released as a substitute to HD 2967, and first variety to be licensed by more than 100 seed companies in India
2015	Improved HD 2932 and Improved HD 2733: First wheat varieties developed through MAS against leaf, stem and stripe rusts

Besides bread - wheat, Indian scientists developed highly resistant durum-wheat varieties, some of them even resistant to *UG 99*, and thus saturated the Central Indian wheat belt with such varieties preventing rust epidemic and protecting main bread-wheat belt of north India by limiting inoculum build-up and spread. Some of these semi-dwarf thermo-insensitive durum lines are exceptionally high yielding (8 tons/ha) even under rain-fed and limited irrigation conditions. To control Karnal bunt, which appeared the main bottleneck in export of wheat and for grain quality, integration of resistance with spraying of fungi-toxicants and cultural practices were standardized for production of practically Karnal bunt-free wheat in high-production areas (Sharma et al., 2004; 2005).

Rice

The semi-dwarf varieties were strengthened in terms of the incorporation of multi-gene resistance against diseases like bacterial blight as well as for quality characteristics in Basmati rice (Joseph et al., 2004). The genetic potential of semi-dwarf rice varieties are being further strengthened with tolerance to increasing abiotic stresses such as submergence, drought, salinity, alkalinity and acidity (Singh et al., 2011). These strategies shall help in developing climate-resilient varieties with sustained high productivity. Moreover, these developments in the case of Basmati rice have greatly boosted quality rice production and export from India from INR 294 CR in 1990-91 to INR 33,000 CR in 2014, and have established it as a niche crop for foot hills of Himalayas in the Indo-Gangetic Plains. This genetic manifestation associated with the identified geographical region has been protected, and is proving a commercial boon to local farmers and the nation (Siddiq et al., 2008).

To break the yield barrier in rice, hybrid rice research was initiated in 1989 through a research network, which has given rise as many as 72 rice hybrids. These include a landmark hybrid Pusa RH 10, the world's first superfine grain aromatic rice hybrid, developed at the IARI (Siddiq et al., 2008). The restorer line of this hybrid, PRR 78 was developed through selective inter-mating of fertile progenies from cross between male sterile-line Pusa 3A with partial restorer Haryana Basmati.

The world's first early-maturing, semi-dwarf Basmati rice variety, Pusa Basmati 1509, with a seed-to-seed maturity of 120 days, was developed through conventional breeding, and released for commercial cultivation in 2013. These developments demonstrate that remarkable achievements could be attained through focused conventional breeding programmes. Through such an approach, per day productivity of Basmati rice varieties has increased from about 14 kg to 50 kg (Singh, 2015). Landmark rice varieties post-Green Revolution are given in Table 1.3.

Table 1.3: Post-Green Revolution landmark rice varieties

Year	Technology/Variety
1988	FR 13A: A pure-line selection from Kalam Banka, served as a donor for submergence tolerance gene "Sub 1"
1989	IR 64: One of the most widely cultivated rice variety introduced from the IRRI, Philippines. This variety is resistant to multiple biotic stresses such as bacterial blight, green leaf hopper, rice tungro virus and grassy stunt virus and moderately resistant to sheath blight and blast
1989	Pusa Basmati 1: The first semi-dwarf, high-yielding and photoperiod insensitive superior quality Basmati rice variety, developed at IARI, New Delhi, which laid the foundation for Basmati rice revolution
1992	Vandana: A semi-tall, photoperiod insensitive variety possessing high degree of drought tolerance. It was released for Bihar and Odisha
1994	Pusa 44: A semi-dwarf, photo-insensitive variety was bred at the IARI, New Delhi, from a cross between IR8 and IARI 5901-21. It was bred for sturdy stem with profuse tillering, suitability for cultivation under high-input conditions. This variety brought a rice revolution in Punjab
1996	KRH 2: A semi-tall hybrid (IR58025A/KMR 3R) developed at the Regional Research Centre at Mandya, one of the most popular rice hybrids in India
1998	ADT43: A semi-dwarf, photo-insensitive, short duration (110 days) variety with yield of 5.9 tons/ha, very popular in Tamil Nadu
2001	Pusa RH 10: World's first superfine grain aromatic rice hybrid, matures in 115 days with yield potential of 8-10 tons/ha, possesses excellent grain and cooking quality traits
2001	PA 6444: Widely cultivated rice hybrid developed by Proagro
2001	Bayer Crop science India hybrid rice Arize 6444 with wider adaptability and consistently higher yield, incidentally private seed sector accounts for more than 99 % of estimated 40,000 tons of hybrid rice seed sown in India and Bayer is global leader
2003	Pusa Basmati 1121: Developed at the IARI, New Delhi, this has exceptionally high cooked kernel length (20 mm), currently grown on 1.2 million ha; contributing 60% of Basmati rice production and export earnings of INR 29,299 CR annually
2007	Improved Pusa Basmati 1 (Pusa 1460): The first product of MAS in rice, pyramiding of genes *xa13* and *Xa21* for resistance to bacterial blight in genetic background of Pusa Basmati 1
2007	Improved Sambha Mahsuri (RP-Bio 226): An improved version of Sambha Mahsuri, developed through pyramiding genes *xa5, xa13 and Xa21* for bacterial blight resistance using MAS
2008	Bayer rice hybrid Arize Dhani, tolerant to BLB
2009	Sahbhagi Dhan: Developed in partnership with the IRRI and CRRI, Cuttack, thus the name Sahbhagi (cooperation); has high degree of drought tolerance
2009	Swarna Sub1: An improved version of Swarna, the mega rice variety, developed through MAS by incorporating *Sub1* gene from FR13A
2014	Pusa Basmati 1509: A semi-dwarf, non-shattering, early (matures in 115-120 days), high- yielding (6.5 tons/ha) Basmati variety with extraordinary kernel length of 9.5 mm developed at the IARI, New Delhi. Saves 5-6 irrigations (33% water saving).This variety surpassed all weaknesses of Pusa Basmati 1121
2017	Bayer rice hybrid Arize AZ 8433 DT, with in-built tolerance to brown plant hopper, the most dreaded insect of rice

Maize

In maize, in late 1980s, in view of global experience and emergence of private sector, national breeding programme was reoriented towards single-cross and other simple hybrids. First three-way (single cross x inbred) hybrid, Trishul, was released in 1991 and first single- cross, 'Paras', in 1995. These were followed with releases of many more single-cross hybrids. Resistance breeding resulted in the development of cold-tolerant maize varieties, namely, Pratap and Pratap 1, to promote cultivation of winter maize (Khera and Dhillon, 1984).

Concentrating on quality improvement quality protein maize (QPM) having o2 gene and modifiers for hard endosperm, popcorn, sweet corn, baby corn, specialized starch (waxy, amylo) and high oil maize, were developed. The work on QPM was initiated 15 years ago with the introduction of QPM lines from the CIMMYT. QPM hybrids rich in essential amino acids, i.e. lysine and tryptophan have been released, namely, Shaktiman 1 and Shaktiman 2 and Vivek Maize QPM 9 (Gupta et al., 2009). Increasing use of hybrid seeds more from the private seed companies have resulted in increasing national maize production from 1.7 million tons in 1950-51 to nearly 24 million tons in 2013-14 with yield increase of 0.5 to 2.5 tons/ha (DMR, 2014).

Pearl-Millet

A large number of genetically diverse male-sterile lines were developed and utilized during 1980-1995 for development of hybrids (Rai et al., 2006). Forty hybrids were released for general cultivation, overcoming downy- mildew problem and doubled the rate of yield enhancement. Later, both seed and pollinator parents were diversified, and another 60 hybrids were released for diverse ecological niches (Yadav et al., 2012). Presently more than 100 hybrids are being marketed by both public and private sectors (more from private sector seed companies), and are currently being grown by farmers over 70% of approximately 10 million ha, increasing national productivity to 1,000 kg/ha.

Sorghum

As regards sorghum, new hybrid production continued with utilization of new CMS sources, and the national average yield of kharif sorghum increased to 1,250 kg/ha in 2012 from 557 kg/ha in 1965. Also, the period saw the development of rabi sorghum varieties both by the All-India Coordinated Sorghum Programme and by various states and private seed companies, increasing yield from 502 kg/ha in 1965 to 935 kg/ha in 2010. Thus, hybrids were found more productive and accepted by the farmers.

Grain-legumes

The concentration for this group of crops during this period has mostly been on breeding for resistance against various biotic and abiotic stresses and developing short-duration, photo-insensitive varieties with desirable plant type with the main focus on sustained and increased pulse production, especially when pulses are being pushed to harsher and marginal growing areas. These efforts have resulted in breeding wilt resistance in chickpea and pigeon pea, sterility mosaic resistance in pigeon pea, powdery-mildew resistance in pea and urd bean, rust resistance in large- seeded varieties of lentil and horse gram, yellow mosaic virus (YMV) resistance in mung bean and bean common mosaic virus (BCMV) resistance in rajmash (dry bean). Monogenic to digenic control to powdery- mildew resistance in pea has been reported. Sharma and associates at the IARI reported single recessive gene, er1er1, being resistant to powdery mildew in pea, and this gene mapped on chromosome 6.

In chickpea, systematic breeding programmes have led to the development of resistant varieties against major diseases, particularly Fusarium wilt and Ascochyta blight. A major breakthrough has occurred in developing bold-seeded kabuli varieties with high-yield potential such as 'KAK 2', 'BG 1003', 'BG1053' and 'JKG 1'. A salt-tolerant variety 'CSG 8962' ('Karnal Chana') has been developed for irrigated areas with moderate salinity, and drought-tolerant genotypes, 'ICCV 10', 'Phule G5', 'K KSO', 'Vijay', have been identified. Development of short- duration and new plant- type varieties have also greatly helped in expanding cultivation of pulses into new agro-ecological niches and cropping systems such as summer mung bean between wheat and rice crops, urd bean in coastal areas and chickpea in rice fallows in south.

In lentil, a hybridization programme of macrosperma x microsperma was initiated at the IARI in1972 with growing macrosperma strains under extended photoperiod through artificial lights or by taking an off-season crop at Lahaul Valley during summer to override difference in flowering time to facilitate crossing. Crosses between microsperma and macrosperma lentils yielded genotypes with 100-seed weight above 5 g coupled with extra earliness, indicating scope for increasing productivity and production of lentil (Sharma et al., 1993).

In pigeon pea, efforts were made for identification of useful genes for yield, pest and disease resistance in wild relatives and transferring them into the cultivated *Cajanus cajan*, especially from *Atylosia platycarpa* and *A. scarabaeoides*. Analysis of inter-generic hybrids between *Cajanus*, *Atylosia* and *Rhynchosia* resolved biosystematic relationships among these genera and helped in visualization of breeding strategies as reported by Pundir and Singh

(1987). Several new high-yielding determinate pigeon pea lines developed through pure-line selection have been reported. They are suitable to multiple cropping under intensive crop production systems.

The development of first pigeon pea hybrid, ICPH 8, released in 1991, based on the genetic-male sterility (GMS) system by the ICRISAT was considered a milestone in the history of pigeon pea breeding (Saxena et al., 1992). Further breakthrough in hybrid pigeon pea was achieved through development of two CMS sources, namely, A2 system (derived from *C. scarabaeoides*; Tikka et al., 1997) and A4 system (derived from *Cajanus cajanifolius*; Saxena et al., 2005). Indian researchers have been recognized for developing first CGMS-based hybrid, GTH 1, in pigeon pea showing improvement up to 25.3% in productivity over the pure-line varieties. The hybrid was released for commercial cultivation in 2004 (Singh, 2015).

Oilseeds

In rapeseed-mustard, following basic breeding strategies, 22 varieties were released in 1980s and 1990s and 41 during the 1st decade of 21st century. The mustard breeders were able to convert fragile, tall and long-duration *Brassica juncea* with a yield of around 300 kg/ha into a robust non-shattering, early-maturing plant type, producing above 1,100 kg on an average yield/ha, recording a three-fold increase in the yield potential, meeting around 20% of the national requirement of edible oilseeds annually (Chopra et al., 1995).

The three-line technology was perfected and hybrid PGHS 51 (Gobhi Sarson, *Brassica napus*), based on *tour* CMS system by the PAU, Ludhiana, in 1994 (Banga et al., 1995) and Hyola 401, based on *pol* CMS from Advanta in 1997/2000 were released. In Indian mustard (*B. juncea*), sustained efforts helped release five CMS-based hybrids, NRCHB 506 (*mori* CMS- based) and DMH 1 (*126-1* CMS- based) in 2009, Coral PAC 432 in 2011, Coral PAC 437 in 2012 and 44 S 01 in 2012 (all *ogu* CMS- based) as reviewed by Singh (2015).

On quality improvement, low erucic acid variety, Pusa Karishma, first low erucic acid Indian mustard, was released for Delhi in 2004, which was followed by several others, namely, Pusa 21, Pusa 22, Pusa 24, RLC 1, RLC 2 of Indian mustard; and double-low varieties, GSC 5, NUDB 26-11, GSC 6 and TERI Uttam Jawahar of Gobhi Sarson. Current efforts are on to combine low erucic acid with low glucosinolate content in Indian mustard and refining agronomic base to improve yield potential of double-low Gobhi Sarson (*Brassica napus*) strains. During this period, 'Bio 902' popularly known as 'Pusa Jaikisan', became the first variety, derived as a somaclonal variant from variety 'Varuna', released in 1994 (Singh, 2015).

In groundnut, after 1980 with the identification of resistant sources for major diseases and insect- pests at the ICRISAT and in the national programme, resistance breeding received strong stimulus and this gave rise to accelerated use of exotic sources including wild species in hybridization programme and release of varieties with multiple resistance, increased yield from794 kg/ha in 1980 to 998 kg/ha in 1994 (Singh and Nigam, 1996). Backcross breeding was also used for incorporating simply inherited traits. For exploitation of wild species in resistance breeding, probable ancestors of cultivated groundnut, *Arachis hypogaea*, were discerned through conventional genome analysis (Singh and Moss, 1982, 1984). This information was used to implement breeding strategies for transfer of genes conferring resistance to various foliar diseases from diploid wild Arachis species to cultivated tetraploid *A. hypogaea* with or without ploidy manipulations (Singh and Moss, 1984; Singh 1986a, 1986b and 1986c), producing inter-specific hybrids, which were backcrossed to recurrent *A. hypogaea* parents and intermittently selfed to regain agronomic traits without diluting resistance. These led to the production of a large number of cultivated groundnut-like inter-specific derivatives. Several of them were registered as sources of diseases resistance or released as varieties as reviewed by Singh (2015).

Cotton

In cotton, large-scale recombination breeding frequently using introductions like, Super Okra, American Nectar-less, Acala Glandless, etc. and their morphological variants continued. Inter-specific hybridization among cultivated varieties and wild relatives, followed by backcrossing was extensively used for introgression of genes conferring desirable traits, leading to the development of a number of high-yielding varieties. Similarly, the hybrid breeding was further strengthened and intensified both in diploid and tetraploid cultivated species. Important tetraploid inter-specific hybrids (*G. hirsutum* x *G. barbadense*) include Varalaxmi, DCH 32 (Jayalaxmi), NHB 12, HB 224, DHB 105, Sruthi and TCHB 213. Four hybrids, namely, GDH 17, GDH 9, DDH 12 and Pha 46, are based on *G. herbaceum* and *G. arboreum*.

Plant Molecular Biology

The plant molecular biology research in India had its roots in plant physiology and biochemistry disciplines, initiated mainly in the institutions in Kolkata and Delhi in1950s (reviewed by Sopory and Maheshwary, 2001). The rise of plant molecular biology in India has been associated with select institutions in Kolkata, Delhi, Allahabad and Varanasi; although presently high-quality research work is being done at several institutes in Pune, Bengaluru, Hyderabad and Chennai. The early pioneering researchers in the area of plant physiology and biochemistry but orientation towards plant molecular biology included S. Ranjan (Allahabad) who had received doctorate training at Cambridge with F. F. Blackman supported by noted scientists Govindjee, Ravinder Kaur and M. M. Laloraya; S. M. Sirkar at Calcutta University (trained at the Imperial College, London), supported by his students S. P. Sen, B. B. Biswas, B. Ghosh and S. Sen-Mandi. Later this group moved to Bose Institute with S. M. Sirkar as Director of the Institute. S. P. Sen benefitted by his visit to C. Leopoldds laboratory at Purdue University and started working on the isolation and mechanism of action of Auxins. Later S. P. Sen moved to Kalyani University and B. B. Biswas carried forward the development of molecular biology activities at the Bose Institute. During the same period, R. N. Singh at the Botany Department of BHU, Varanasi, also pioneered research work in the area of cell- free nitrogen fixation in blue- green algae. B. B. Biswas from Calcutta University was also attracted to work at BHU and published papers in Nature before moving back to Kolkata to join Bose Institute, where he had a clear lead in the area of plant molecular biology in India.

In the recent years, three major groups have contributed to the understanding of molecular biology. Akhilesh Tyagi and his group at Delhi University had initiated studies on the characterization of the chloroplast genome of Arabidopsis, rice and Vigna.

The other major group working on chloroplastic genes is from the ICGEB, Delhi. Mukherjee and his group have developed a partially purified in-vitro system for chloroplast replication and characterized factors needed for it. Biswas and his student Majumdar at the Bose Institute worked on myoinositol metabolism. Biswas contribution was in establishing and elucidating metabolic cycle involving myoinositol phosphate in germinating and developing seeds. This is one of the major contributions in plant biochemistry from India, and is at present part of text books of plant biochemistry.

Later on major contributions were by Sastry and Rajashekharan from Indian Institute of Science, Bangalore where it was established that in many crop plants cytoplasmic male sterility (CMS) is associated with dysfunction of mitochondria. Sane and his colleagues at the NBRI undertook detailed studies on the biochemical and molecular basis of male sterility using well-known male- sterile restorer and maintainer lines of sorghum and rice. On the basis of their work on the CMS in sorghum, they proposed a model for explaining male sterility in A1 cytoplasm (*Milo-Kafir* combination).

UNO supported establishment of International Centre for Genetic Engineering and Biotechnology (ICGEB) at New Delhi, which is now adequately funded by the DBT, Government of India. Another group at the ICGEB headed by NarendraTuteja presented the first direct evidence for the role of a pea DNA helicase in salinity stress tolerance and pea heteromeric G-proteins in salinity and heat tolerance. Maheswari and Guha at Delhi University developed techniques for induction of haploids in rice for the first time in 1964. (Guha and Maheswari, 1964).

At Delhi University, P. Maheswari had setup School of Experimental Botany that gave outstanding results in the area of Plant Embryology. Earlier during his career at Dacca University, P. Maheshwari came in contact with the leading plant embryologists and tissue culturists of the world, including F. Scoog (Madison), P. White (Bar Harbor), J. van Overbeek (Modesto), E. W. Sinnnot and A. Galston (Yale). He was a great teacher of Plant Physiology and Biochemistry. In 1957, he organized the first international seminar on Modern Trends in Plant Physiology. Due to wide contacts of P. Maheshwari, many students were able to participate in classic research of that time; for example discovery of RNA polymerases in plants and initial evidence for separate genome in chloroplast.

With rapid developments in the area of recombinant DNA technology internationally during the nineteen hundred seventies and eighties, a new Department of Plant Molecular Biology (DPMB) was established at the Delhi University, South Campus, to focus on the frontier areas of research. Akhilesh Tyagi succeeded in cloning and sequencing gene for a 33 kD a manganese binding protein of photosystem II from spinach. The DPMB, which originated in the School of Experimental Botany later became one of main centres of plant genomics in India which together with the ICAR-National Research Centre on Plant Biotechnology (NRCPB) sequenced long- arm of rice chromosome 11 in collaboration with the International Rice Genome Sequencing Project (IRGSP), the first genome sequencing work in India for any plant or animal.

The Indian Agricultural Research Institute (IARI), New Delhi, has been the premier institution under the ICAR where research in the frontier areas of biochemistry, physiology and molecular biology of crop plants were initiated, which subsequently spread to other institutions. At the IARI, S. K. Sinha and S. L. Mehta were the pioneering researchers. During the same time, K. R. Koundal after obtaining training in Don Boulter's laboratory in England, organized cloned seed lectin cDNA and genes from different leguminous plants and also organized a series of short to medium-term training courses on techniques of molecular biology at the NRCPB, New Delhi.

At JNU, New Delhi, Asis Dutta established a separate centre for Plant Genomic Research which was later converted to a full-fledged National

Institute of Plant Genomic Research (NIPGR). Asis Datta was among the first to have cloned and patented plant genes in India, namely Ama1 protein from Amarathus. In 1990s, he started working on improving nutritional quality of potato by transferring Ama1 gene from Amaranthus and also of tomato by transferring oxalate decarboxylase gene from a fungus which he had cloned in his lab. Later, in 1990 Department of Biotechnology, Government of India, established seven Centres of Plant Molecular Biology (CPMB) at JNU, Delhi; Delhi University (DU) South Campus, New Delhi; National Botanical Research Institute (NBRI) Lucknow; Bose Institute (BI), Kolkata; Osmania University (OU), Hyderabad; Madurai Kamraj University (MKU), Maduari; and Tamil Nadu Agricultural University (TNAU), Coimbatore, to intensify research in plant molecular biology in the country.

Among the state agricultural universities, Pantnagar, was the first one to establish a Biochemistry Department under the headship of K. G. Golakota, and was later headed by G. K. Garg, who was one of the first to set-up a Department of Molecular Biology and Genetic Engineering in the College of Basic Sciences and Humanities at Pantnagar in 1990. Genes and promoters of wheat-seed storage proteins were cloned at Pantnagar by Nagendra Singh with the support of a DBT-funded project on the wheat-quality improvement. He also for the first time in India standardized system for transformation and regeneration in wheat using micro-projectile bombardment of wheat embryos with test gene constructs.

Among the CSIR laboratories, outstanding work in the area of plant biochemistry and molecular biology was done at the CFTRI, Mysore, NCL, Pune and NBRI, Lucknow. N. K. Singh worked extensively on biochemistry and molecular biology of wheat - seed storage proteins and identified novel subunits of wheat glutenin and a safe method for destaining of SDS-PAGE protein gels using sodium chloride solutions. P. K. Ranjekar working at the National Chemical Laboratory, Pune, was among the first to start characterization of plant repeat DNA by dissociation ad re-association studies, and later his colleague Vidya Gupta developed molecular markers in papaya and rice.

P. V. Sane who had training with S. Zalik in Canada and R. Park in the USA along with his colleagues P. Nath, R. Tuli and D. V. Amla could set-up one of the six DBT-sponsored national centres of Plant Molecular Biology at the NBRI, Lucknow, and worked on the characterization of poplar chloroplast genome. Rakesh Tuli (NBRI, Lucknow) synthesized codon optimized BT gene, which had potential for commercial applications. The Indian Institute of Science, Bengaluru, is another premier institution where plant biochemistry and molecular biology work was conducted, mostly in the Department of

Biochemistry. A landmark contribution was cloning and transcription of rice histone genes by J. D. Padayatty and co-workers, who published their finding in Nature. This was one of the first works from India using advanced plant molecular biology techniques. In the more recent years, Usha Vijayaraghavan working in the Department of Microbiology and Cell Biology had done outstanding work in the plant differentiation, particularly on homeotic genes controlling leaf and flower development in rice and Arabidopsis. For details of work and the original references, the readers are advised to refer to the review by Singh (2015).

Genomics and Molecular Breeding

Decoding of Crop-Plant Genomes

Genome is defined as the complete set of genetic information present in a single haploid cell of the organism, including its protoplasm. However, decoding of complete genomes has become possible only since the last two decades with the advent of automated DNA sequencing and developments in high-power computing. Two independent techniques of DNA sequencing were developed by Maxam and Gilbert (1977) and Sanger et al. (1977). Both the techniques were made possible with the developments in analytical electrophoresis, allowing separation of DNA molecules differing in size by single nucleotide. While the method of Maxam and Gilbert was based on base-specific chemical cleavage of purified single stranded DNA molecules, Sanger's method relied on enzymatic synthesis of DNA in the presence of di-deoxynucleotides, which generated DNA sequence ladder by random chain terminations at specific bases. It required much smaller amounts of single or double stranded template DNA and became method of choice for the next 30 years with various modifications until the development of three different next generation sequencing (NGS) technologies in 2008, which did away with the need of cloning DNA fragments in bacteria and relied on in-situ PCR amplification of several thousand DNA fragments allowing massive parallel sequencing. The latest, third generation sequencing technologies, e.g. PacBioSMRT sequencing is able to sequence single molecules of DNA without even need of PCR amplification, hence are suitable for studying epigenetic DNA modifications in addition to primary nucleotide sequencing (http://www.pacificbiosciences.com).

The path-breaking human genome project (HGP) was initiated quite early in October 1990 at the National Institute of Health (NIH), USA, under the leadership of James Watson but it was a herculean task and was expected to take fifteen years for its completion because of the huge size of human genome; measuring about three billion base pairs. A National Human Genome Research Institute (NHGRI) was created at the NIH in 1989. The first draft of human genome was published in 2001, but a more complete genome was published in April 2003, two years ahead of the anticipated time. It coincided with the 50th anniversary of the elucidation of double helix structure of DNA by Francis Crick and James Watson in 1953. The HGP was a collaboration of the NHGRI, Office of the Biological and Environmental Research at the Department of Energy, USA, as well as the universities and research centres in countries like Canada, China, France, Germany, Japan, New Zealand and UK. The objective of the HGP was not only to decipher the complete DNA sequence but also to know the entire repertoire of proteins produced within a human being and understanding expression of genes which determine observable traits. Despite the well- established scientific-technological infrastructure and human resource; India did not join this historical initiative and as a consequence was left out of this project.

The first plant genome to be sequenced completely was that of model plant *Arabidopsis thaliana* (The Arabidopsis Genome Initiative 2000). It was the third multi-cellular organism genome sequenced after that of nematode *C. elegans* and fruit-fly *Drosophila melanogaster* (Adams et al., 2000). It was claimed at that time that *Arabidopsis* genome sequence created the potential for direct and efficient access to a much deeper understanding of the plant development and environmental responses, and permitted structure and dynamics of plant genomes to be assessed and understood. A noteworthy aspiration was to determine functions of all *Arabidopsis* genes by 2010; a goal still not fully realized but it has had a catalytic effect on the research community and on how the plant science research is to be conducted.

India missed the opportunity of joining the first international projects for the sequencing of model plant *Arabidopsis thaliana* and human genome but with growing international interest in the area of genomics a National Centre for Plant Genome Research (NCPGR) was set-up in November 1997 with the funding support from the Department of Biotechnology, Government of India, at the Jawahar Lal Nehru University (JNU) Campus, New Delhi. Asis Datta who was leading this centre envisioned that it would play a leading role in plant genome research in India. The centre was upgraded to the National Institute of Plant Genome Research (NIPGR) in November 2002 with Asis Datta as the founding Director. Meanwhile, working from the interim premises of the NCPGR at JNU, Asis Datta and his students started sequencing of expressed sequence tags (ESTs) and identification of new genes and promoters for plant-pathogen interaction and nutritional quality improvement. The NIPGR has indeed lived up to its expectations under the leadership of Akhilesh Tyagi who along with Nagendra Singh from the ICAR-NRCPB was the lead Indian scientists in decoding rice genome. The NIPGR has also been able to generate the draft sequence of a desi type chickpea variety ICC 4958 in-house without any international support (Jain et al., 2013). A Biotechnology Centre was established in 1985 by the Ministry of Agriculture in the ICAR-IARI, New Delhi, under the leadership of V. L. Chopra to carry out advance research in Plant Biotechnology. The Biotechnology Centre was upgraded in 1993 to National Research Centre on Plant Biotechnology (NRCPB) with V. L. Chopra as its Founding Director. It is a premier institute engaged in research of national importance, focusing on major thrust areas such as exploitation of tissue-culture generated somaclonal variations, genetic engineering for biotic and abiotic stress resistance, heterosis for enhancement of Brassica productivity, molecular breeding to unlock genetic potential of diverse germplasm and integrate it with breeding of crop varieties and genome sequencing to latest high throughput Omics sciences, which has provided accessibility of techniques to the Indian scientists at several levels from genome to transcriptome to proteome to metabolome. 'Pusa Jaikisan' mustard was the first product of Biotechnology released by the Centre for commercial cultivation by the Indian farmers. The Centre was also partner in the release of 'Improved Pusa Basmati 1' by the IARI, the first product of the marker-assisted selection by the Indian scientists.

The ICAR-NRCPB has been strengthened with modern equipment and the latest infrastructure to serve as state-of-the-art National Research Centre. It has taken lead and contributed substantially towards plant genomics under the leadership of its successive Directors, R. P. Sharma, K. R. Koundal, P. Anand Kumar and Nagendra Kumar Singh (acting Director). The NRCPB made path-breaking contributions by decoding genomes of rice, tomato, flax and wheat through international collaboration, as well as chickpea rhizobium and pigeon pea genomes in-house under the leadership of Nagendra Singh who joined the Centre in 2000 to lead the Genomics and Molecular Breeding group. Publication of pigeon pea genome by the Centre in 2011 marked an important milestone in the Crop Science research in India. It was the first genome of any species sequenced entirely in India and the first pulse-legume crop genome sequenced anywhere in the world (Singh et al., 2012).

Due to availability of modern equipment leading to fast and automated system of genome sequencing, genome of 30 plant species has been sequenced. These are (1) *Arabidopsis thaliana* 'Columbia', (2) *Oryza sativa* L. *indica* '93-11', (3) *Japonica*, (4) *Japonica* 'Nipponbare', (4) *Populus trichocarpa* Poplar 'Nisqually1' (5) *Vitis vinifera* Grapevine 'PN 40024' (5) *Carica papaya* L. Papaya 'Sun Up', (6) *Lotus japonicus* 'Miyakojima MG20', (6) *Zea mays* L. Maize 'B 73', (7) *Sorghum bicolor* (L.) Moench'BT x 623', (8), *Cucumis sativus* L. Cucumber' Chinese Long 9930', (9) *Brachypodium distachyon* wild, (10) *Glycine max* cv. 'Williams 82', (11) *Malus domestica* Borkh.

Apple 'Golden Delicious', (12) *Theobroma cacao* L. Cocoa 'B97-61/B2', (13) *Medicago tancatula* Barrel medic, (14) *B. rapa* mustard Chiifu-401-42, (15) *Fragaria vesca* Strawberry 'Hawai 4', (16) *Hordeum vulgare* L. Barley 'Morex', (17) *Solanum tuberosum* L. Potato DH Phureja DM1-3, (18) *Cajanus cajan* L. Millsp Pigeon pea 'Asha', (19) *Linum usitatissimum* Flax 'CDC Bethune', (19) *Musa acuminata* Banana 'DH-Pahang', (20) *Solanum lycopersicum* Tomato 'Heinz 1706', (21) *Cicer arietinum* Chickpea Desi 'ICC4958', (22) *Cicer arietinum* Chickpea Kabuli 'CDC Frontier', (23) *Citrus sinensis* Sweet Orange 'Valentia', (24) *Prunus persica* Peach DH 'Lovell', (25) *Coffea canephora* Coffee DH, (26) Wheat cv.'Chinese Spring' *Triticum aestivum.*

For tomato, potato, India has been a partner and genome sequencing of pigeon pea and chick pea has been done by Indian scientists themselves without involvement of any other agency. Pigeon pea genome sequencing has been done by Nagendra Singh and the team at IARI, New Delhi and that of chick pea by Rajeev Varsheny and the team at ICRISAT, Hyderabad (as per compilation by Nagendra Singh and included in the review by Singh (2015).

The ICAR-NRCPB, NIPGR and Delhi University South Campus (DUSC) have been the pioneers of plant genome research in India. Apart from these, Punjab Agricultural University (PAU), Ludhiana, under the leadership of Kuldeep Singh played a key role in decoding wheat genome as part of the International Wheat Genome Sequencing Consortium (IWGSC), whereas ICAR-Central Potato Research Institute (CPRI), Shimla, was part of the Potato Genome Sequencing Consortium. Availability of high-quality reference sequence of plant genomes is paving way for fast-track discovery of genes for important agronomic traits in many field and horticultural crops. So far, India has its name in seven different crop genome publications. The ICAR-NRCPB contributed to five of these genomes, except for potato genome which was partnered by the ICAR-CPRI and chickpea genome which was sequenced entirely in India at the NIPGR (Jain et al., 2013). The first draft of pigeon pea genome was sequenced entirely in India at the ICAR-NRCPB (Singh et al., 2012). International Rice Genome Sequencing Project India's journey in the area of genomics began truly with her joining the International Rice Genome Sequencing Project (IRGSP) in 2000. Rice was the first and the most important food crop whose genome was decoded under the leadership of Japan. The International Rice Genome Sequencing Project (IRGSP) consortium was formed in 1998 for completing a high quality reference genome of Japonica rice variety 'Nipponbare' with average base accuracy of 99.99% in the span of ten years. The IRGSP was a consortium of publicly funded laboratories in ten-member countries—Japan, USA, China, Taiwan, France, India, Thailand,

Korea, Brazil, and the UK. It adopted a clone-by-clone shotgun sequencing strategy so that each sequenced BAC clone could be associated with a specific position on the genetic map and adhered to policy of immediate release of the sequence data to the public domain.

With the availability of rice genome sequence data in the public domain and creation of desired computational infrastructure and human resource expertise through the IIRGS project, Nagendra Singh and Tilak Raj Sharma embarked upon creating complete database of all rice genes. It was felt necessary to create information resource for use by rice geneticists and breeders for gene mapping and crop improvement through molecular breeding. The 'Vanshanudhan' rice gene database is used extensively at the NRCPB for identification of new QTLs and genes. The unique feature of this database is that it is user-friendly, and it links gene identification code and gene function in an integrated manner.

Tomato Genome Sequencing Consortium

Sequencing of tomato genome began in 2004 by an international consortium of laboratories in ten countries including China, France, India, Japan, Korea, Netherlands, Spain, Italy, UK and USA. The initial approach was to sequence only gene-rich euchromatic region representing about twenty per cent of genome using a BAC-by-BAC approach, and more than 1,200 BAC clones were sequenced by 2008. However, in 2008 a whole-genome shotgun sequencing approach based on second generation sequencing technologies was adopted, which in conjunction with the earlier data yielded high quality assemblies of complete tomato genome including heterochromatic regions. The International Tomato Annotation Group (ITAG) annotated the genome builds generated by this combined sequencing approach with active participation from India. A high quality reference genome of tomato was published in 2012 with 24 Indian co-authors (TTGC, 2012). The period witnessed revolutions in DNA sequencing technologies with the introduction of 454 pyro-sequencing, ABI's SOLiD sequencing and Solexa-Illumina's reversible chain termination technology, together called NGS technologies. With these there was a paradigm shift from clone by clone sequencing to whole genome shotgun sequencing of large genomes requiring high level computational and bioinformatics capabilities.

Potato Genome Sequencing Consortium

Potato is the important non-grain food crop and is central to global food security in India and globally. Potato genome is highly heterozygous, and to counter this problem a homozygous doubled-monoploid potato clone was used to sequence and assemble 93% of the 844-megabase potato genome by the Potato Genome Sequencing Consortium (TPGSC, 2011). Total 39,031 protein-coding genes were predicted and evidence was provided for at least two genome duplication events suggesting a palaeopolyploid origin of the potato. The consortium identified 2,642 genes specific to this large angiosperm clade asterid. They also sequenced a heterozygous diploid clone and showed that it has large number of gene presence/absence variants and potentially deleterious mutations, which could be a likely cause of inbreeding depression. The ICAR-Central Potato Research Institute (CPRI), Shimla, was partner in this effort with S. K. Chakrabarti as Principal Investigator and V. U. Patil as co-author of the paper published in Nature for their involvement in experimental design, data generation and data analysis. The financial support to the potato genome project at the CPRI, Shimla, was provided by the ICAR through its XI Plan EFC grant. A genome laboratory was created with the installation of the second generation sequencing facility Illumina's GAII sequencer.

Pigeonpea Genomics Initiative

Pigeonpea (*Cajanus cajan* L. Millsp.) is the second most important pulse-legume crop of India grown and consumed in all parts of the country. Pigeon pea genomics initiative (PGI) was started by the ICAR under the Indo-US Agricultural Knowledge Initiative (AKI) in November 2006. Nagendra Singh played a crucial role in the development and coordination of the PGI network as the National Facilitator for the Biotechnology component of the Indo-US AKI. The objective was to create a genomic and expressed sequence resource rich enough for genetic mapping of traits of agronomic importance. Now pigeon pea has moved from an orphan legume to one where genomics-assisted breeding for a sustainable crop improvement will become a reality (Varshney et al., 2010).

The main objectives of the project were: (i) development of at least 100, 000 pigeon pea ESTs and EST based SSR/SNP markers; (ii) development of genomic SSR markers; (iii) development of mutant lines and mapping populations as resource for gene discovery; (iv) construction of a high density reference molecular linkage map; (v) identification of markers and genes for important agronomic traits; (vi) development and upkeep of a pigeonpea genome informatics platform; and (vii) sequencing gene-rich BAC clones of pigeonpea. After termination of AKI-PGI in 2009-10, the work continued with the support from ICAR-NPTC project and coordination by Nagendra Singh at the ICAR-NRCPB, which led to the publication of first draft of pigeon pea genome of variety 'Asha' using long reads of FLX-454 pyrosequencing identifying 47,004 protein coding genes (Singh et al., 2012). Immediately after the publication of pigeon pea genome by India, a group of international scientists coordinated by Rajiv Varshney from the ICRISAT published another draft of the pigeonpea variety 'Asha' based on the short reads Illumina sequencing at the BGI, China (Varshney et al., 2012).

Chickpea Genome Initiative

Chickpea (*Cicer arietinum* L.) is the third most important food-legume crop in the world and most important pulse-legume crop of India. Immediately after completion of the tomato genome sequencing, the scientists at the NIPGR, New Delhi, started working on decoding of chickpea genome using FLX-454 sequencing technology. They produced the first draft genome sequence of a desi-type chickpea genotype IC 4958 using next-generation sequencing platforms, bacterial artificial chromosome end sequences and a genetic map (Jain et al., 2013). A 520 Mb assembly was generated that covered 70% of the predicted 740-Mb genome length, and more than 80% of the gene space. Genome analysis predicted 27,571 protein coding genes and 210 Mb as repeat elements. The gene expression analysis performed using 274 million RNA-Seq reads identified several tissue-specific and stress-responsive genes. Although segmental duplicated blocks were observed, the chickpea genome did not exhibit any indication of recent whole-genome duplication.

The ICRISAT produced a draft Kabuli type Canadian chickpea variety 'CDC Frontier' in the same year through international collaboration using the BGI platform (Varshney et al., 2013b).

International Wheat Genome Sequencing Consortium

Its genome size is huge, six-fold larger than the human genome and forty-fold larger than the rice genome, requiring a huge international collaboration for decoding. Immediately after the successful completion of rice genome and publication of landmark reference rice genome paper in 2005, the ICAR-NRCPB was approached by Bikram Gill, noted wheat Geneticist from the Kansas State University USA, for joining the International Wheat Genome Sequencing Consortium (IWGSC). A detailed network project proposal was conceptualized and submitted to the ICAR-DBT in 2005 for decoding complex wheat genome in collaboration with the IWGSC. India was able to join late in 2010 with the DBT funding support, and took responsibility of sequencing wheat chromosome 2A. The Indian wheat genome network involves ICAR-NRCPB and DU South Campus from New Delhi and PAU Ludhiana. The ICAR-NRCPB under the leadership of Nagendra Singh, DUSC under the leadership of J. P. Khurana and PAU, Ludhiana, under the leadership of Kuldeep Singh, who is also the overall coordinator of the Indian project, took the responsibility of sequencing chromosome 2A.

This allowed establishment of Plant Genome laboratory in India at the PAU, Ludhiana, and the first in a SAU with generous support from the DBT. Flow cytometry sorted chromosome arm specific genomic DNAs, and BAC libraries were made available by Jaroslav Dolzel of the Czech Republic to each member of the IWGSC. A chromosome based draft sequence of wheat genome was published in Science (July 2014) with four Indian scientists as co-authors (TIWGSC, 2014). On 17 August 2018, the IWGSC published in the international Journal Science a detailed description and an analysis of the reference sequence of the bread wheat genome, the world's most widely cultivated crop. The research article authored by more than 200 scientists from 73 research institutions in 20 countries presents the DNA sequence of the bread wheat variety Chinese Spring ordered along the 21 chromosomes (www.wheatgenome.org).

With efficient technologies like Illumina and Nanopore, genome sequencing (whole genome sequencing, targeted sequencing, exome sequencing, amplicon sequencing etc.) has become routine which generates lots of data but without an able bioinformatics tools and precision phenotyping, these sequences remain only a large database of limited use. Even if a complete genome sequence of one cultivar is available, still another cultivar has a part of sequence which is unique to it and quite possible that this part of genome provides its specialty. Therefore one needs to sequence and phenotype all the germplasm being used in the crossing programme and stored in the genebank. With genome sequencing, one can have now genome wide SNP markers, copy number variation (CNV) and presence/absence variation (PAV) which will led to greater possibility of tight trait-marker association and ultimately for marker assisted selection. Another good use of genome sequencing is going to be in gene editing where a specific nucleotide can be edited through Crispr/CASs system for trait specific improvement (Shiv Agrawal, ICARDA, personal communication).

Marker-Assisted Breeding of Crop Varieties

Marker-assisted selection (MAS) is gaining momentum in transferring simply inherited traits with minimum linkage drag and maximum recurrent parent genome recovery. Nevertheless, the complete potential of MAS has not been availed in India owing to lack of high throughput genotyping and phenotyping facilities and paucity of precise molecular markers to track the target traits. The most successful use of MAS in India has been in rice which is described as follows.

Rice Success Story Based on MAS

Marker-assisted backcross breeding (MABB) is the most feasible approach for improving elite rice varieties for traits such as disease and insect resistance, tolerance to salinity and drought etc. To improve the most widely cultivated Basmati rice variety Pusa Basmati 1 for bacterial blight resistance, the resistance genes *xa13* and *Xa21* were incorporated through MABB and the country's first marker- assisted selection bred rice variety 'Improved Pusa Basmati 1, resistant to BB was released during 2007 (Joseph et al., 2004; Gopalakrishnan et al., 2008). Similarly, Improved Samba Mahsuri was developed for BB resistance by incorporating genes *xa5*, *xa13* and *Xa21* into the genetic background of elite rice variety Samba Mahsuri (Sundaram et al., 2008). The isogenic lines Pusa 1592 and Pusa 1612 of high yielding extra-long slender aromatic rice variety Pusa Sugandh 5 were improved for BB and blast resistance by incorporating *xa13* and *Xa21*; and *Piz5* and *Pi54*, using the donor parents Improved Pusa Basmati 1; and C 101A51 and Tetep and were released during 2012 and 2013 respectively. Similarly, the parental lines of aromatic rice hybrid Pusa RH 10, i.e. Pusa 6B and PRR 78 were improved for BB and blast resistance by incorporating resistant genes *xa13* and *Xa21*; and *Piz5* and *Pi54* using the donor parent Improved Pusa Basmati 1 (Basavaraj et al., 2010) and for blast resistance by incorporating the resistance genes using the donor parents C 101A51 and Tetep, respectively (Singh et al., 2012a, 2013b). Several other elite genotypes improved for bacterial blight, blast and sheath blight resistance in rice are presented in Table 1.4 (Singh, et al., 2015).

Table 1.4: Improvement of elite rice varieties for various diseases through marker-assisted breeding

Recurrent parent	Donor parent	Trait	Gene	References
PR 106	IRBB 59	BB	*xa5+xa13+Xa21*	Singh et al. (2001)
Pusa Basmati 1	IRBB 55	BB	*xa13+Xa21*	Joseph et al. (2004), Gopalakrishnan et al. (2008)
Samba Mahsuri	SS 1113	BB	*xa5+xa13+Xa21*	Sundaram et al. (2008)
PRR 78 and Pusa 6B	Pusa 1460	BB	*xa13+Xa21*	Basavaraj et al. (2010)
Type 3	SS 1113	BB	*xa13+Xa21+sd1*	Rajpurohit et al. (2010)
Basmati 370 and Basmati 386	SS 1113	BB	*xa5+xa13+Xa21*	Bhatia et al. (2011)
Basmati 370 and Taraori Basmati	Improved Sambha Mahsuri	BB	*xa13+Xa21*	Pandey et al. (2013)
PRR 78	Tetep and C101A51	Blast	*Piz5+Pi54*	Singh et al. (2012a)

Recurrent parent	Donor parent	Trait	Gene	References
Pusa 6B	Tetep	Blast and Sheath Blight	*Pi54+qSBR11-1 + qSBR11-2 + qSBR7-1*	Singh et al. (2014)
Improved Pusa Basmati 1	Tetep and Sheath	BB, Blast	*xa13+Xa21 + Pi54 + qSBR11-1* Blight	Singh et al. (2012b)
ADT43	IR BB60	BB	*xa5+xa13+Xa21*	Perumalsamy et al. (2010)
Triguna	PR 106	BB	*xa5+xa13+Xa21*	Sundaram et al. (2009)

There are examples of successful application of MAS in wheat for transfer of leaf rust resistance gene; in chickpea for introgression of gens for resistance to fusarium wilt; in soybean for transfer of null allele of KTI (Kunitz Tripsin Inhibitor) and in maize for developing QPM (quality protein maize) hybrids through transfer of opaque 2 (*o2* gene). For details, Singh et al. (2015) may be referred.

Basic Plant Breeding Concepts and Procedures - Most Relevant to Commercial Plant Breeding

The above given historical perspectives on history of selected crop breeding in India indicates dominance of following three issues.

1. Practicing selection (both mass and pure-line) in early phase of any breeding programme.
2. In self-pollinated crops, selection of complementary parental cultivars, making crosses and handing of segregating generations mostly through pedigree method of breeding or its modifications.
3. Introgression of few genes of importance into parental or commercial cultivars using back-cross method of breeding, sometimes facilitated by marker assisted selection.
4. Hybrid breeding wherever possible.

The relevant operational plant breeding concepts and related breeding procedures commonly used in commercial plant breeding are briefly described as follows (Ram, 2020).

Selection, Heritability and Genetic Advance

In 1859 Darwin proposed in The Origin of Species that natural selection is the mechanism of evolution. Darwin's proposition was that the adaptation of populations to their environments resulted from natural selection and that if this process continued for long enough, it would ultimately result into the origin of new species. Darwin's 'Theory of Evolution through Natural Selection' presumed that plants change gradually by natural selection operating on variable populations and this was the outstanding discovery of the 19th century with direct relevance to plant breeding.

Effective selection, the oldest method of plant breeding practiced by the early plant breeders depends on the existence of genetic variability which in a specific breeding population depends on the germplasm included in the breeding population and its selection history. When genetic variability in a breeding population is insufficient to permit attainment of a specific goal, it will be necessary to increase the variability by using either mutagenic treatment or introduction of new germplasm. Collections of germplasm are available for most crop species in the national gene banks and those of CGIAR funded international agriculture centres, like IRRI, CIMMYT, ICARDA and ICRISAT, etc and these can be used to increase the available genetic variability, either by crossing them with the currently available germplasm or by selecting within the germplasm for local conditions. Obviously, genetic variation is fundamental for selection, by which progress in plant breeding can be made.

Heritability

Heritability is usually defined as the proportion of the total or phenotypic variance for a given trait that is strictly due to genetic variation. Heritability estimates serve two purposes in plant breeding. First, they show the relative ease with which different traits are selected under a given testing regime. The other purpose for heritability estimates is prediction of selection progress. The change in population mean due to selection is a function of heritability (h^2) and selection differential S.

Dudley and Moll (1969) pointed out that each estimate of heritability in plant breeding relates specifically to the reference population of genotypes for which it was estimated. Also, each estimate relates only to the particular set of environments and experimental conditions in which the genotypes were tested. Thus, it is difficult to generalize heritability estimate from one population of genotypes to another or from one set of experimental conditions (for example, plot size) to another. When interpreting or comparing heritability estimates, it is of course necessary to note the unit on which they are based (for example, plot mean, family mean, etc).

Response to Selection

Genetic variation forms the basis for selection in plant breeding. Selection results in the differential reproduction of genotypes in a population so that gene frequencies change and, with them, genotypic and phenotypic values (mean and variance) of the trait being selected. Response to selection, or genetic advance in one generation of selection, is measured by the difference between the selected population and their offspring population, which is denoted as R. Response to selection has been referred to by several different names, including genetic progress, genetic advance, genetic gain, and predicted progress or gain, and has been generally denoted as R.

The selected individuals have a mean x, while the offspring of the selected population has a mean y. The difference between the selected population and the original population is defined as the selection differential, and denoted by S, i.e.

$$S = \bar{x} - \mu$$

The response to selection, R, can be written as

$$R = \bar{y} - \mu$$

The relationship between S and R is determined by heritability, and is expressed as follows.

$$R = h^2S$$

$$= k\sigma_p h^2$$

where k = the standardized selection differential read from the table based on the percentage of the plants selected and σ_p is the phenotypic standard deviation.

How much of the selection differential is realized in the offspring population depends on the heritability of the trait. The heritability, h^2, in the formula can be either h_N^2 or h^2_B (depending on whether the offspring are produced by sexual or asexual reproduction, respectively). From the above formula,

$$\bar{y} = \mu + h^2S$$

The population mean of the offspring derived from the selected individuals is equal to the parental population mean plus the response to selection. When $h^2 = 1$, the selection differential will be fully realized in the offspring population so that its mean will deviate from the parental population by S. When $h^2 = 0$, the selection differential cannot be realized so the offspring population mean will regress to its parental population. When $0 < h^2 < 1$, the selection differential is partially realized so that the mean of the offspring population will deviate from the parental population by h^2S. It is very useful to predict the response before selection is undertaken (Xu, 2010). Diagrammatic representation of response to selection/genetic advance is depicted in Fig. 1.1 (Xu, 2010).

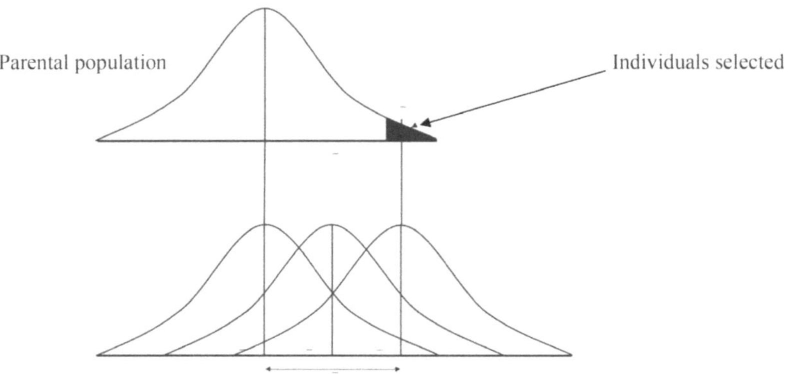

$h^2 = 0, 0.5, 1.0$

Fig. 1.1: Distribution of parental and progeny populations with a selection intensity of 5%, because the phenotypic values of the selected plants include both a genetic and an environmental component, the progeny means depend on the heritability of the trait selected (Xu, 2010)

In most plant breeding programmes, more than one trait needs to be improved simultaneously at a time. For example, a high-yielding cultivar susceptible to a prevalent disease would be of little use to a grower. Recognition that improvement of one trait may cause improvement or deterioration in associated traits serves to emphasize the need for the simultaneous consideration of all traits which are economically important in a crop species. Three selection methods, which are recognized as appropriate for the simultaneous improvement of two or more traits in a breeding programme, are index selection, independent culling, and tandem selection.

Tandem Selection

In this mode of selection, selection is focused on one trait at a time. Once this is achieved, selection for another trait is exercised and so on. Thus, this selection is also known as sequential selection or serial improvement. This takes a long time and is effective when genetic correlation does not exist between the target traits. This is hardly practiced now.

Independent Culling

Independent culling requires the establishment of minimum levels of merit for each trait. An individual with a phenotype value below the critical culling level for any trait will be removed from the population. That is, only individuals meeting requirements for all traits will be selected. In independent culling level, one might select for all the characters at the same time but independently, rejecting all individuals that fail to come up to a certain standard for each character regardless of their values for any other of the characters.

Index Selection

A selection index is a single score which reflects the merits and demerits of all target traits. Selection among individuals is based on the relative values of the index scores. Selection indices provide one method for improving multiple traits in a breeding programme. The use of a selection index in plant breeding was originally proposed by Smith (1936) who acknowledged critical input from Fisher (1936). Subsequently, methods of developing selection indices were modified, subjected to critical evaluation, and compared to other methods of multiple trait selection.

It has been shown, however, that the most rapid improvement of economic value is expected from selection applied simultaneously to all the component characters together, appropriate economic weight being given to its relative economic importance, its heritability, and the genetic and phenotypic correlation between different characters (Falconer, 1960). Practice of selection for economic value is thus a matter of some complexity. The component characters have to be combined together into a score or index, in such a way that selection applied to the index, as if the index were a single character, will yield the most rapid improvement of economic value. It is generally recognized that a selection index is a linear function of observable phenotypic values of different traits. There are a number of forms of the equations available from index selection for multiple traits in grain. To construct a selection index, the observed value of each trait is weighted by an index coefficient,

$$I = b_1x_1 + b_2x_2 + \ldots + b_nx_n$$

where I is an index of merit of an individual, x_i represents the observed phenotypic value of the ith trait, and $b_1 \ldots b_n$ are weights assigned to phenotypic trait measurements represented as $x_1 \ldots x_n$. The b values are the products of the inverse of the phenotypic variance–covariance matrix, genotypic variance–covariance matrix, and a vector of economic weights. A number of variations of this index, most changing the manner of computing the b values, have been developed. These include the base index of Williams (1962), the desired gain index of Pesek and Baker (1969), and retrospective indexes proposed by Bernardo (1991). The emphasis in the retrospective index developments is on quantifying the knowledge experienced breeders have obtained. Baker (1986) summarized all select indexes in plant breeding developed before that time. Selection index is hardly applied in applied crop breeding programmes.

Even though theoretically superior, index selection has had limited use in plant breeding. One reason for this is the difficulty of assigning relative weights to the traits under selection. In most crops this problem can be simplified by giving yield a weight of 1 and all other traits a weight of zero. This utilizes direct and indirect selection together.

Population Improvement

Several different 'population breeding methods' can be used, for example, (1) bulk; (2) mass selection; and (3) recurrent selection. One of the methods used for managing large populations of segregates was the 'bulk method' proposed by Harlan and his team consisting of Harlan himself, M. L. Martini and H. Stevens in 1940 for multi-parent crosses. This concept changed the breeding methodologies for self-pollinated species. Mass selection is a system of breeding in which seeds from individuals selected on the basis of phenotype are bulked and used to grow the next generation. Mass selection is the oldest breeding method for plant improvement and was employed by early farmers for the development of cultivated species from their ancestral forms. This method is still used in breeding programmes having less resources and field facilities and in principle is quite effective for the characters having high heritability.

The enhancement or improvement of open-pollinated populations of crops such as rye, maize and sugar beet, herbage grasses, legumes, and tropical trees such as cacao, coconut, oil palm, and some rubber, depends essentially on changing the gene frequencies so that the favourable alleles are fixed, while maintaining a high (but far from maximal) degree of heterozygosity. Recurrent selection is a method of plant breeding associated with quantitatively inherited traits by which the frequencies of favourable genes are increased in populations of plants. The methodology is cyclical with each cycle encompassing two phases: (i) selection of genotypes that possess the favourable or required genes; and (ii) crossing among the selected genotypes. This leads to a gradual increase in the frequencies of the desired alleles. While recurrent selection is often successful, it also has potential limitations in closed populations and this has led to numerous modifications and alternative schemes (Hallauer and Miranda, 1988). Recurrent selection breeding methods have been applied to a wide range of plant species, including self-pollinated crops.

Recurrent Selection

Recurrent selection in its broadest sense is any cyclical scheme of plant selection by which frequencies of favourable genes are increased in plant population. Some form of recurrent selection has always been used in plant improvement, but the methods of selection often were not used systematically. Recurrent selection methods were developed primarily for the improvement of discrete classes; quantitatively inherited traits, which involve a large number of genetic factor pairs, each with a small effect, and genotypes cannot be classified into discrete classes. Additionally, environmental effects tend to obscure the genetic effects. Quantitative traits are measured metrically and are used to determine the effects of selection. Gene frequencies are not known, but if selection is effective, gene or genotypic frequencies must be changing in the desired direction. The basic objective of all recurrent selection methods is to increase the frequency of desirable genes in a population so that opportunities to extract superior genotypes are enhanced (Hallauer, 1981).

Recurrent selection can be broadly defined as the systematic selection of desirable individuals from a population followed by recombination of the selected individuals to form a new population. The basic feature of recurrent selection methods is that they are procedures conducted in a repetitive manner, in cycle one after another, including development of a base population with which to begin selection, evaluation of individuals from the population, and selection of superior individuals as parents that can be crossed to produce a new population for the next cycle of selection.

After each cycle of recurrent selection a new population is formed. The initial population that is developed for a recurrent selection programme is referred to as the base, or cycle 0, population. The population formed after one cycle of selection is called the cycle 1 population; the cycle 2 population is developed from the second cycle of selection, and so on.

Recurrent selection procedures are conducted for primarily quantitatively inherited traits. The objective of recurrent selection is to improve the mean performance of a population of plants by increasing the frequency of favourable alleles in a consistent manner in order to enhance the value of the population and to maintain the genetic variability present in the population as effectively as possible. In addition, separation of the genetic and environmental effects is an important facet of effective recurrent selection methods. The improved populations can be used as a cultivar per se, as parents of a cultivar-cross hybrid and as a source of superior individuals that can be used as inbred lines, pure-line cultivars, clonal cultivars, or parents of a synthetic line. Successful recurrent selection results in an improved population that is superior to the original population in mean performance and in the performance of the best individuals within it. Ideally, the population will be improved without its genetic variability being significantly reduced so that additional selection and improvement can occur in the future. Recurrent selection is complementary to inbred development procedures; in fact the concept of recurrent selection was developed, particularly for out-crossing crops, to rectify limitations in inbred development by continuous selfing that rapidly leads to inbreeding and allele fixation and thus inadequate opportunity for selection. First, recurrent selection increases the frequency of favourable alleles in the base population by repeated cycles of selection. Secondly, recurrent selection maintains the degree of genetic variation in the population to allow sustained progress from subsequent cycles of selection. Genetic variation is maintained by recombining a sufficiently large number of individuals to reduce random fluctuations in allele frequencies, i.e. genetic drift (Xu, 2010).

Recurrent selection as used in corn, has the following operational features: (1) plants from a heterozygous sources are self-pollinated and at the same time are evaluated for some desirable character or characters; (2) plants with inferior performance for the character or characters under improvement are discarded; (3) the superior plants are propagated from the selfed seed; (4) all possible intercrosses among these superior progenies are made by hand, or , if this is impractical, the intercrosses are made by open pollination among the selected progenies and; (5) the resulting intercross population serves as source material for additional cycles of selection and intercrossing (Allard, 1960). According to Hull (1952), as reproduced by Allard (1960), "Recurrent selection was meant to include reselection generation after generation, with interbreeding of selects to provide for genetic recombination. Thus, selection among isolates, inbred lines, or clones is not recurrent selection until selects are interbred and new cycle of selection is initiated".

Since the late 1950s, extensive research has been conducted to determine the relative importance of different genetic effects on the inheritance of quantitative traits for most cultivated plant species. Quantitative genetic research has provided extensive information to assist plant breeders in developing breeding and selection strategies (Hallauer, 2007).

Four types of recurrent selection have been recognized based on the way in which plants with desirable attributes are identified. These four types are: (1) simple recurrent selection, (2) recurrent selection for general combining ability, (3) recurrent selection for specific combining ability and (4) reciprocal recurrent selection. In simple recurrent selection, plants are divided into a group to be discarded and a group to be propagated further on the basis of phenotypic scores taken on individual plants or their selfed progeny. Since test crosses are not made, the effective use of simple recurrent selection is restricted to characters with sufficiently high heritability where superior plants can be identified visually. The remaining three methods of recurrent selection differ from simple recurrent selection in the sense that test crosses are made to measure the combining ability. If the tester used is of broad genetic base, variations in performance in a group of test crosses will primarily be due to differences in general combining ability and this method will be known as recurrent selection for general combing ability. In recurrent selection for specific combining ability, the tester used is an inbred line with narrow genetic base. In this case, variation in the test-cross performance is attributed to differences in specific combining ability. Reciprocal recurrent selection uses two heterozygous source populations, each of which is a tester for the other. This method, with certain limitations, provides selection for both general and specific combining ability.

Different methods of recurrent selection (recurrent selection for general combining ability, recurrent selection for specific combining ability and reciprocal recurrent selection) have following implications with respect to gene action involved:

(1) If dominance is incomplete, recurrent selection for general combining ability and reciprocal recurrent selection are about equal to each other, and both are superior to recurrent selection for specific combining ability.

(2) If dominance is complete, all the three methods (recurrent selection for general combining ability, recurrent selection for specific combining ability and reciprocal recurrent selection) have equal efficiency.

(3) If dominance is over-dominance, recurrent selection for specific combining ability and reciprocal recurrent selection are equal in efficiency and both are superior to recurrent selection for general combining ability.

Pure-Line Breeding Procedure

The steps involved, in general, are as follows.

1. **Year 1:** Have a variable base population (new introduction, segregating population from crosses, landrace) and space plant the same under appropriate growing conditions. Harvest better looking plants individually and put the clean seed into individual bags/packets without mixing the seed of any two plants.

2. **Year 2:** Grow individual plant-to-progeny rows from the seed of plants harvested in first year. Harvest better and uniform looking progenies individually and bulk the seed from such selected and harvested progenies separately into individual bags. Ensure no mixing of seed from any two progenies harvested. Each bulk is termed as a new strain/breeding line.
3. **Year 3-5:** Conduct station trial of newly created strains/breeding lines along with the standard checks in replicated trial and record all desirable traits.

Year 6-9: Conduct coordinated multiplication trials for the lines selected based on station trial for three years at multi-locations for detailed observations and identification of a few outstanding lines for further release and notification provided the selected line(s) are found to be better than the currently grown best cultivar.

Pedigree Selection in Self-Pollinated Crops

This method has been first described by H. H. Lowe in 1927. This is the most widely used method to handle segregating generations following crosses between two parents to develop new cultivars combining desirable traits from both the parents. In this method, superior types are selected in successive segregating generations and a record is maintained of all parent-progeny relationships. The records help the breeder to advance only progeny lines with plants that exhibit genes for the desired traits. Selection begins in F_2 generation where individual plants are selected which in judgment of breeder will produce the best progeny in F_5/F_6 when the selection is finally concluded. In F_3 and F_4 generations, many loci will have become homozygous and family characteristics begin to appear. By F_5/F_6 generation, most families are expected to be homozygous and uniform and hence selection within such families is no longer effective. At this stage selection emphasis shifts to the selection among families rather than within families.

Selection of parents is of great significance for the success of this method. Normally, this method is used to replace already established variety. Therefore, one parent should be selected on the basis of proven performance in the particular area where the breeding is in progress. The second parent should be such that it complements the first parent for the character under improvement. Sometimes, three or four parents may also be used to supply the desired character by way of three-way or double crosses.

Year-wise, Steps Under Pedigree Selection

The procedures are as follows.

Year 1 (crossing): Identification of desired parents and making sufficient crosses to have about 50-100 crossed seeds/parental combination.

Year 2 (F_1): Growing 50 F_1 plants including parents to authenticate hybridity and harvesting of F_2 seeds in bulk from F_1 plants.

Year 3 (F_2): Approximately 5000 F_2 plants in cereals are space planted to ensure full expression of the characters and evaluation of individual F_2 plants. F_2 population provides first opportunity to start selection. Obviously, undesirable plants are often removed as soon as they are identified. It is here that beginning plant breeder will need to develop a ruthless-ness toward unpromising plants. About 300 desirable F_2 plants are selected and harvested separately keeping records of their identities. From these 300 plants, about 50 plants are discarded after laboratory examination for seed size, shape, colour, etc. Thus, there are 250 F_3 seed bags/packets each having come from separate individual F_2 plants. Seed from any two packets should not be mixed. Parental/ check cultivars should also be planted at appropriate intervals.

Year 4 (F_3): Seed from selected individual F_2 plants are progeny rowed. There are approximately 250 individual progeny rows/cross. F_3 generation is the beginning of line formation. It is desirable to plant check cultivars for comparison and to select the superior rows and superior plants within superior rows. The progeny rows are space planted for easy visual selection and record keeping. F_3 family row should contain large number of plants (25-30) to permit the true family features to be evident so that most desirable plant(s) can be selected. Selection at this stage is both within and between rows by first identifying superior rows and selecting 3-5 plants from each selected progeny row to plant the next generation. In total 125 plants are selected from 50 superior rows.

Year 5 (F_4): Selected F_3 plants are planted in plant-to-row fashion as in F_3 generation in 125 rows. The progenies become more homogeneous (homozygosity is 87.50 %). Distinct lines start appearing in F_4 generation. Therefore, selection is exercised more between rows rather than within rows. In total about 90 superior individual plants are selected from selected 40 progeny rows. Families that are distinctly inferior should be discarded while more than one plant may be selected from exceptional and outstanding families. However, generally, number of plants selected and advanced does not exceed the number of F_3 families.

Year 6 (F_5): Plants selected in F_4 numbering 90 are grown in 90 rows in F_5. These 90 rows are subjected to rigorous selection keeping the end objectives in mind and about 35 best performing progeny rows are selected. In total about 80 individual plants are selected from the short-listed 35 progenies. Depending upon seed requirement for trial in the next generation, selected plants within progeny rows can be bulked to have sufficient seeds for trial.

Year 7 (F_6): 80 breeding lines as constituted from F_5 are planted and evaluated in initial station trial along with other breeding lines pooled from other crosses and checks in a randomized block design with two replications. Each trial normally has 20 genotypes (18 new breeding lines and two checks). From each trial, breeding lines found to be superior to the checks are selected for further evaluation in next year in station advanced varietal trial.

Year 8 (F_7): Superior lines from each initial station trial are pooled together for evaluation in station advanced varietal trial where each trial normally has 20 new breeding lines (16 new lines plus 4 checks). This trial is planted in randomized block design with four replications. Superior new breeding lines (about 4)/cross are short-listed for further evaluation under coordinated multi-location trials.

Year 9-11 (F_8-F_{10}): The selected lines are entered into national coordinated varietal trials where there are yield evaluation for three years and then the most outstanding line is identified for released and ultimately the release proposal is submitted to Central Sub-Committee on Crop Standards, Notification and Release of Varieties (Field Crops) for their consideration and release and notification so that the product enters into formal seed production and distribution chain. The number of plants/lines planted and selected will vary from crop to crop and will be influenced by the availability of land and other resources at the breeding station and also the technical input of the breeder.

The pedigree method is not suited for inbreeding in greenhouses where the expression of agronomic characters is not typical of that observed under field conditions in the area to which the genotypes are adapted. The length of time required for cultivar development when only one generation is grown each ear for pedigree selection generally exceeds that of methods that can take advantage of multiple generations each year. This limits the current use of the pedigree method for cultivar development in certain crop breeding programmes.

Single Seed Descent Method in Self-Pollinated Crops

The method of single seed descent (SSD) was born out of a need to speed up the breeding programme by rapidly inbreeding a population prior to initiating individual plant selection and evaluation while reducing a loss of genotypes during the segregating generations. The concept as per Acquaah (2007) was first proposed by C. H. Goulden in 1941 when he obtained the F_6 generation in two years by reducing the number of generations grown from a plant to one or two while conducting multiple plantings per year using greenhouse and off-season planting. H. W. Johnson and R. L. Bernard described the procedure of harvesting a single seed per plant for soybean in 1962. However, Brim (1966) suggested modified pedigree method i.e. single seed descent (SSD) as an economical way to develop homozygous lines and simultaneously preserve the genetic variation in the population. Single seed descent in the strict sense refers to planting a segregating population, harvesting a sample of one seed per plant, and use one-seed sample to plant the next generation. When the population has been advanced from F_2 to the desired level of inbreeding, the plants from which lines are derived will each trace to different F_2 individuals. The number of plants in a population declines each generation due to failure of some seeds to germinate or some plants to produce at least one seed. As a result not all of the F_2 plants originally sampled in the population will be represented by a progeny when generation advance is completed. When the SSD is used, most breeders prefer to harvest and retain a reserve sample of seed each generation in case the one planted is destroyed. In SSD, only one seed is chosen at random from each plant for further use. Since the number of plants to be grown is considerably reduced by such a procedure, the individual generations can be grown in greenhouses where two to three generations per year can be achieved.

This method allows the breeder to advance the maximum number of F_2 plants through F_5 generation. This is achieved by advancing one randomly selected seed per plant through early segregating generations. The focus on the early stages of the procedure is on attaining homozygosity as fast as possible and without selection. Discriminating/selection among plants starts after attainment of homozygosity. This method is best suited for small grain crops as well as legumes that can tolerate close planting and produce some seeds per plant. The species that can be forced to mature rapidly are more suited for breeding by SSD method. The handling of plants/lines starting from F_6 and onward is as given in case of pedigree selection. The classical procedure of SSD has sometimes been modified and practiced as single pod/multiple seed method where 2-4 seeds per plant are taken and only half of the bulk seed thus produced is used to advance the generations.

Back-Cross Breeding in Self-Pollinated Crops

The application of back-cross breeding first proposed by H. V. Harlan and M. N. Pope in 1922 provides a precise way of improving varieties that excel in a large number of attributes but are deficient in a few key characteristics. F. N. Briggs too started an extensive back-crossing programme to develop bunt resistant varieties of wheat in 1922. As the name implies, the back-cross method of breeding makes use of a series of back-crosses to the variety to be improved during which the character or characters in which improvement is sought is maintained by selection. At the end of back-crossing, the gene (or genes) being transferred will be heterozygous. Selfing at the end of last back-cross produces homozygosity for this gene pair and coupled with selection will result in a variety similar to the well adapted and recurrent parent but superior for the particular characteristic for which improvement programme was initiated.

In back-cross method of breeding, the adapted and highly desirable parent is called the recurrent parent and the source of desirable trait being transferred is called as donor parent. It is most effective and easy to apply when the missing trait is qualitatively inherited, dominant and produces a phenotype that is readily observed in a hybrid plant. The procedure for transferring a recessive trait is similar to that for dominant trait but involves an additional step of selfing. Back-crossing is also used to transfer entire sets of chromosomes in a foreign cytoplasm to create a cytoplasmic male sterile (CMS) genotype as has been practiced in maize, onion and wheat, etc. This is done by crossing the donor (of chromosomes) as male until all donor chromosome are recovered in the cytoplasm of the recurrent parent. Back-cross breeding has also been used to develop isogenic lines (lines that differ only in alleles at a specific locus) for traits like disease resistance and plant height, etc.

Transfer of Dominant Gene by Back-Crossing

The generalized steps in back-cross breeding for a dominant trait are given in Fig. 1.2 (Acquaah, 2007). With the availability of markers, the foreground and background selection can be made accurate and now it is possible to reduce the number of years of back-crossing substantially.

In general, the steps for transferring a dominant gene are straightforward. Following the first cross between the parents, phenotypic selection is adequate for selecting plants that exhibit the target trait. Recessive genotypes are discarded. The recurrent parent traits are not selected at this stage. The next cross is between selected F_1 and recurrent parent. This step is repeated for several cycles. After satisfactory recovery of the recurrent parent, the selected plant in BC_5/BC_6 will be homozygous for other alleles but heterozygous for the desired traits. The last back-cross is followed by selfing to stabilize the desired gene in the homozygous state. All homozygous recessive segregates are discarded.

Year	Generation	Crosses	Action
1	P	A (susceptible) x B (resistant) rr x RR	Select parents, make 10-20 crosses, harvest F_1 seed
2	F_1	↓ Rr (resistant) x rr	Grow F_1 plants and back-cross F_1 to recurrent susceptible parent
3	BC_1F_1	↓ rr Rr x rr	Discard susceptible plants (rr); back-cross Rr (about 30-50 heterozygous plants resembling with the recurrent) with rr
4	BC_2F_1	↓ rr Rr x rr	Discard susceptible plants (rr); back-cross Rr (about 30-50 heterozygous plants resembling the recurrent parent) with rr
5	BC_3F_1	↓ rr Rr	Grow BC_3F_1
6	BC_3F_2	↓ rr rr Rr RR	Discard susceptible plants (rr); progeny row
7	BC_3F_3	↓ rr Rr RR RR	Select BC_3F_3 progenies with resistance and high yield
8		↓ RR x rr	Back-cross superior lines to rr
9	BC_4F_1	↓ Rr x rr	
10	BC_5F_1	↓ rr Rr x rr	Discard susceptible plants; back-cross resistant plants to rr, BC_5 should closely resemble the recurrent parent
11	BC_6F_1	↓ Rr rr	Self BC_6, select 300-400 desirable plants and harvest them individually
12	BC_6F_2	↓ rr Rr RR	Discard susceptible (rr) and grow progeny rows
13	BC_6F_3	↓ rr Rr RR RR	Discard susceptible (rr) progenies and select true breeding resistant progenies (RR), conduct yield test of the back-cross derived line with the recurrent parent to determine equivalence before release.

Fig. 1.2: Generalized steps in transferring a dominant gene in back-cross breeding

Heterosis and Hybrid Breeding

Charles Darwin in 1877 showed that crosses of related strains did not exhibit the vigour of hybrids. He observed heterosis, i.e. the tendency of cross-bred individuals to show qualities superior to those of both parents, in crops like maize and concluded that cross-fertilization was generally beneficial and self-fertilization injurious. In 1879, William Beal demonstrated hybrid vigour in maize by using two unrelated cultivars. The best combinations yielded 50% more than the mean of the parents. Reports by Sanborn in 1890 and McClure in 1892 confirmed Beal's earlier reports and extended the generality of the superiority of hybrids over the average of the parental forms (Xu, 2010). Now exploitation of heterosis through hybrid breeding has become a routine activity in most of the crop breeding programmes including both self and cross-pollinated crops. There are two theories to explain heterosis.

Dominance theory first proposed by Davenport in 1908 was later elaborated by Lerner. The second theory known as over-dominance theory was first proposed by Shull in 1908 and later by Mather and Jinks.

Dominance Theory

This theory assumes that hybrid vigour in plants is conditioned by dominant alleles and recessive alleles being deleterious or neutral in effect. It follows then that a genotype with more number of dominant alleles will be more vigorous than the one with lesser number of dominant alleles. Crossing two parents with complementary dominant alleles will lead to concentration of more number of dominant favourable alleles in the F_1 hybrid than the either parent. This is more favoured than the over-dominance theory by most of the scientists, although neither is completely satisfactory. Based on this theory, it should be possible to isolate true breeding inbred lines having all the desirable dominant genes together in homozygous condition and this line should be as vigorous as the F_1 hybrid. However, this does not occur and it is hypothesized that linkage and large number of genes are the major hurdles in developing inbred lines that contain all homozygous dominant alleles. Inbreeding depression is observed as the deleterious recessive alleles that are suppressed in the heterozygous condition, are expressed in the homozygous condition. However, lately highly productive inbred lines in maize have been developed paving the way for developing and commercializing single cross hybrids in maize. The dominance theory of heterosis is illustrated with the following example where each dominant genotype contributes two units and a recessive genotype contributes one unit to the phenotype. The example shows that F_1 is better than the better parent and shows heterosis.

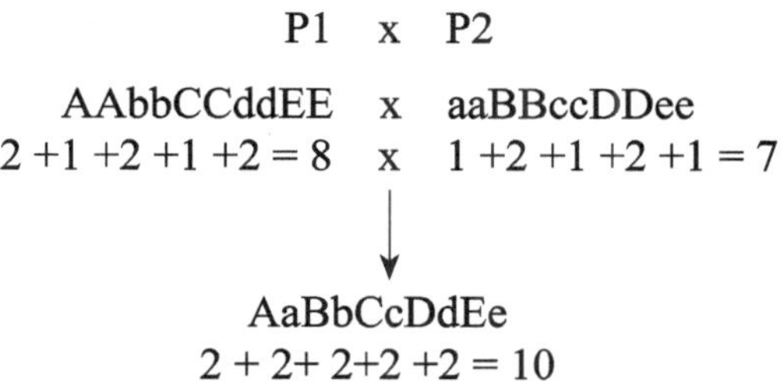

D. S. Falconer developed a mathematical relationship between heterosis, degree of dominance and difference in gene frequency in the parents of the cross. The equation is as follows.

HF1 = $\sum dy^2$ where

HF1 = Deviation of hybrid performance from the mid-parent value.

d = Degree of dominance.

y = Difference in the gene frequency in the parents involved in the cross.

Maximum heterosis will occur when the values of the two variables (d and y) are each unity. This means that the parents to be crossed are fixed for opposite alleles (y = 1.0) and there is complete dominance (d = 1.0).

Over-Dominance Theory

The first theory of heterosis (heterozygosity theory of heterosis) was proposed by Shull and East independently in 1908. It was restated later by Shull in 1848 as "Hybridity itself - the union of unlike elements, the state of being heterozygous – has........a stimulating effect upon the physiological activities of the organisms whose effect disappears as rapidly as continuous inbreeding reduces the progenies to the homozygous types". For this, he proposed the term "heterosis". According to this theory, heterozygote is superior to either homozygote i.e. Aa is always superior to AA or aa. Initially this theory was applicable to single locus but later on it was extended to a series of alleles where the increased heterosis was in the order as a1a2 < a1a3 < a1a4, etc.

Numerically, over-dominance theory of heterosis can be demonstrated as follows, if we assume that the recessive, heterozygote, and homozygote dominants contribute one, three and two units to the phenotypic value, respectively. Heterozygosity per se is most superior of the three genotypes (Ram, 2020).

	P1 (aabbCCDD)	x	P2 (AABBccdd)
Phenotypic value	1 + 1 + 2 + 2 = 6	↓	2 + 2 + 1 + 1 = 6
		AaBbCcDd 3 + 3 + 3 + 3 = 12	

Male Sterility

Commercial exploitation of heterosis or hybrid vigour depends upon ease of producing hybrid seed in bulk and distributing the same to the farmers in the desired quantity. This is where male sterility, self-incompatibility, apomixis and chemical hybridizing agents come to our mind. East and Mangelsdorf (1925) reported for the first time gametophytic self-incompatibility and later, Gerstel (1950), Hughes and Babcock (1950), and Mather (1950) reported sporophytic self-incompatibility. Jones and Davis (1944) discovered cytoplasmic-genetic (C) type of male sterility in onion. The discovery of cytoplasmic male sterility in maize by Rhoades (1933) provided the basis needed for the subsequent production of hybrids in sorghum, and other plants. In maize, different cytoplasmic male sterility sources, namely, T, C and S-CMS (Rhoades, 1933; Duvick, 1959) were used to produce hybrid seeds. In sun flower, the genetic male sterility, cytoplasmic male sterility and fertility restorer genes were discovered in 1965, 1969 and 1970, respectively. Leelercq (1966) as mentioned by Kharkwal and Roy (2004) reported the discovery of cytoplasmic male sterility in the progeny of inter-specific crosses of sunflower. Meyer and Meyer (1965) created cytoplasmic male sterility by placing *hirsutum* cotton genome into the cytoplasm of wild species (*G. anomalum*). In pearl millet, Tift- 23A which is of US origin is the source of male sterility. In India, hybrid seed production is maize is commonly done using manual detaselling and pollination. However, in sorghum, pearl millet and sunflower, use of cytoplasmic genetic male sterility is common mostly in the private sector seed companies.

In pigeon pea (an often cross-pollinated crop), a hybrid (ICPH-8) was developed by K. B. Saxena and his team at ICRISAT, Hyderabad using genetic male sterility system. Similarly, self-incompatibility has been used to produce hybrids in several cruciferous crops. In rice, hybrids based on three line system (cytoplasmic genetic male sterility, A, B, C or A, B, R lines) have been developed and commercialized in China, India, Vietnam, Egypt and several other countries. Two line rice hybrids based on environmental (temperature and photo-period) sensitive make sterility have also been developed in China. The first super-fine grained, aromatic rice hybrid, Pusa Rice Hybrid-10 developed at Indian Agricultural Research Institute has been released for commercial cultivation in India in 2001 (Kharkwal and Roy, 2004).

In conclusion, the predominant system of hybrid seed production on commercial scale in India mostly by the private seed companies is as follows.

Manual Method: Maize

CMS Using Three Lines System: Sorghum, pearl millet, sunflower, rice, mustard

GMS: Pigeon pea but could not make the desired impact

Marker Assisted-Selection (MAS)

Selecting plants having desirable traits in a segregating generation is an important and critical task in any crop breeding programme. In this exercise, the plant breeder has to deal with hundreds and thousands of individual plants. Close association between molecular markers and target traits has been demonstrated in many cases. With the aid of molecular marker technology, one can now identify, locate and estimate the genetic effects of individual quantitative trait loci (QTL) as reported by Dudley (1993). Although a number of scientists have proposed models for estimating gene effects and other parameters, the interval mapping approach developed by Lander and Bostein (1989) is recognized as the pioneer work in the field of QTL mapping. With the advent of molecular markers, such as hybridization based restriction fragment length polymorphism (RFLP) and polymerase chain reaction (PCR) based RAPDs, AFLPs, SNPs, mini- and micro-satellites (Williams et al., 1990; Vos et al., 1995) plant breeders can now not only fingerprint the genotypes, but can also practice marker-assisted selection.

Marker-assisted selection/marker-assisted breeding/marker-aided selection may greatly increase the efficiency and effectiveness of the selection compared to normal phenotypic selection in any conventional breeding programme. Once markers that are tightly linked to genes or QTL of interest have been indentified, breeders may use the markers as diagnostic tool to indentify the plants carrying the genes/QTLs prior to field evaluation.

MAS for early generation selection is commonly practiced for disease resistance where a susceptible parent is crossed with the resistant parent and F_1 plants are self-pollinated to produce a F_2 population. By using a marker to assist selection, plant breeder may substitute large field trials and eliminate many unwanted genotypes and retain only those plants that possess the desirable genotype based on the presence of tightly liked marker. 75 % of the plants may be eliminated after one cycle of MAS for a recessive disease resistance gene from the F_2 population. This is important because breeders typically use very large population consisting of about 2000 F_2 plants derived from a single cross and may use populations derived from hundreds of crosses per year (Ram, 2020).

Advantages of MAS

- Time saving from substitution of complex field trials with molecular tests.
- Elimination of unreliable phenotypic evaluation associated with field trials due to environmental effects.
- Selection of genotypes at seedling stage.
- Gene 'pyramiding' combining multiple genes simultaneously.
- Avoiding the transfer of undesirable or deleterious genes (linkage drag) during introgression of genes from wild species.
- Effective selection of traits with low heritability.

Limitations of MAS

- MAS results may not be easily published.
- Reliability and accuracy of QTL mapping.
- A loose association may exist between marker and gene/QTL.
- There are limited markers and limited polymorphism for markers in breeding material.
- Effect of the genetic background.
- QTL x environment interactions.
- High cost of MAS.
- Application gap between research laboratories and plant breeding institutes.
- Knowledge gap between plant breeders and molecular biologists or scientists in other disciplines.

Marker-Assisted Back-Crossing (MAB)

Backcross breeding is a traditional breeding method routinely used for introgression of a desirable trait from a donor to a recipient variety which is otherwise suitable but for the trait under introgression. The target traits are introgressed from a donor genotype into the genomic background of an elite variety (recipient genotype) through backcrossing. Traditional backcrossing programmes are planned on the assumption that the proportion of the recurrent parent genome is recovered at a rate of $1 - (1/2)^{n+1}$ for each of n generations of backcrossing. This means that usually six to seven generations are required to achieve with > 99% genetic similarity to the recurrent parent. However, the use of markers can accelerate backcross breeding through the precise transfer of genomic regions involved in the expression of target traits (foreground selection) and by speeding up the recovery of the recurrent parent genome (background selection). This method, described as marker-assisted backcrossing (MAB), is particularly useful for pyramiding genes or QTL for resistance against a pathogen or pest and for traits that are highly influenced by the environment. In India, using this approach has been practiced for disease resistance breeding in rice, wheat, and chickpea and for developing Kunitz Tripsin Inhibitor free lines in soybean and for developing quality protein maize (QPM).

When the donor is an un-adapted or wild genotype having several undesirable genes also, the desirable genes can be introgressed through simple backcross breeding using MAS. This method helps select rare progenies having only targeted regions from a donor parent using linked markers (foreground selection) and rest of the genome from a recipient parent using whole genome markers (background selection). The use of markers that flank a target gene (<5cM on either side) can minimize the linkage drag which is common in conventional breeding methods. The indirect selection of desirable plants using MAS, as discussed, helps to save time by reducing the number of backcross generations (3 - 4 generations) and also increases genetic grain. MAS may also improve mass selection and increase efficiency through progeny testing and decreasing the number of replications and increasing selection intensity. Marker-assisted backcross breeding has been used successfully to incorporate genes or QTL for both qualitative and quantitative traits in a number of crop species and in some cases leading to the development of improved cultivars.

Machine Learning in Plant Breeding

Technological developments have revolutionized measurements on plant genotypes and phenotypes, leading to routine production of large and complex data sets. This has led to increased efforts to extract meaning from these measurements and to integrate various data sets. Concurrently, machine learning is rapidly evolving and is now widely applied in science in general and in plant genotyping and phenotyping in particular. In many disciplines in modern plant science, the study of genomes and genotypes plays an important role in precision breeding. The DNA sequencing revolution has allowed breeders to determine the full genomes of many plants, including model organisms such as *Arabidopsis thaliana*, scientifically interesting species of flowering plants, trees, algae and mosses, and economically important crops such as rice, maize, soybean, cotton, and wheat. Genetic variation – from single-nucleotide polymorphisms (SNPs) and small insertion/deletions to gene copy number variation and genome structural variation – can likewise easily be measured, leading to the availability of population-wide genotype collections for many species (Torkamaneh et al., 2018). Similar developments allow the scientists to routinely measure genome-wide differences at the biochemical level, i.e. in concentrations and interactions of molecules in the cell such as RNA, proteins, and metabolites, yielding so-called "omics" data sets. More recently, high-throughput, automated measurements of macroscopic phenotypes or traits have become possible, i.e., quantification of the development, morphology, growth, or yield of plant tissues, organs, whole plants, or canopies, as well as of the environments in which plants grow (Zhao et al., 2019).

As reviewed by van Dijk et al. (2021), in the resulting "big data" era in plant sciences, a challenge in both fundamental and applied research (e.g. breeding applications) is to explain or predict phenotypes from the underlying genotypes under different environmental conditions (Fig. 1.3, van Dijk et al., 2021). Genotypic variation leads to differences in the biochemical makeup of cells, which in turn are linked with the environment influence, organ formation, plant growth, and eventually traits relevant in agriculture, such as yield and tolerance to biotic and abiotic stresses. Unraveling the effects of genotypic variation and environment on phenotypes yields fundamental insights into the regulation of important processes in plant development and growth and the ability to predict yield and quality traits from genotypes in specific environments, which is essential in modern molecular plant breeding/precision breeding. Analyzing phenotypes measured at these different levels or linking these phenotypes to genotypes increasingly calls for processing and integration of large, noisy, and heterogeneous data sets. Machine learning (ML), a set of computational approaches to find predictive patterns in data, plays an increasingly important role in these efforts (van Dijk et al., 2021)). In various scientific and engineering domains, ML has driven a spur of recent innovations, and it is set to do the same in plant research, particularly plant breeding (Ma et al., 2014).

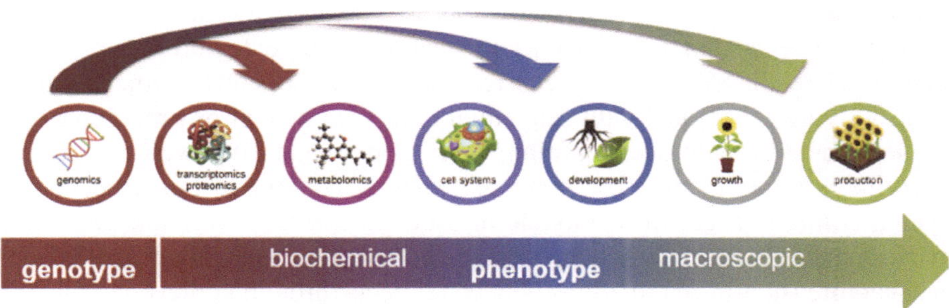

Fig. 1.3: Linkage of variation at the genotype level (genomics data) to phenotypic variation at various biochemical levels of organization (-omics data), at the cellular level, and at the macroscopic level at different scales (van Dijk et al., 2021)

Machine learning (ML) develops algorithms that learn to perform specific tasks based on a provided data set. It is a subfield of artificial intelligence that is widely used in research and the industry. A major distinction is between supervised and unsupervised learning. Supervised learning tasks aim to predict an output (either a discrete label, in the case of classification, or a numerical value, in the case of regression) for a given object, given a set of input features that describe the object. To this end, supervised learning optimizes a predictive model by fitting its parameters to perform well on labeled training data, consisting of inputs and corresponding known outputs. The resulting models can then make predictions for new, unseen test data. Care should be taken to avoid over fitting, situations in which the model performs well on training data but does not generalize well to unseen data. Unsupervised learning, on the other hand, seeks patterns in unlabeled data. Compared to supervised approaches, it is less straightforward to quantify whether an unsupervised model performs well; since unsupervised approaches do not consider outputs, typically there is no such thing as "training data" for these approaches. However, the patterns found by unsupervised approaches can support interpretation of large data and can be useful to prepare data for more successful supervised learning, by allowing researchers to focus on relevant patterns. For both supervised and unsupervised machine learning, a plethora of algorithms have been developed, each with their own strengths and weaknesses.

In recent years, deep learning (DL) has had spectacular success in specific application areas. This approach involves using networks containing "neurons", connected to each other in such a way that signals can be transmitted through the network. Weights involved in the calculation of the signals are adjusted during model training. Typically, neurons are aggregated into layers, and deep learning involves using several of such layers. Different types of artificial neural networks exist, each adapted to specific aspects of input data. Convolutional neural networks are very effective at image analysis; they preserve spatial relationships by learning image features using small subsets of neighboring pixels. Recurrent neural networks on the other hand are especially useful to deal with sequential data such as text or time-dependent signals. In addition to deep learning, other modern and often well-performing algorithms include ensemble methods, such as random forests or boosting, which combine a group of models in order to improve over the prediction performance of each individual model, and support vector machines, which can capture (nonlinear) relations between objects through kernels (calculating similarity between different objects).

In applications, data often need to be pre-processed depending on their type and the choice of algorithm, and care should be taken what measurements are taken or calculated to best represent the objects. The most decisive factor in successful application of ML is the availability of well measured and, in particular for supervised approaches, correctly labeled training data. The choice of model is important as well: simpler models are less likely to overfit but cannot always model more complex relations between input features. The appropriate level of model complexity for a given data set is however difficult to decide a priori. In practice, it is useful to start with a simple model and if needed for the data at hand increase model complexity. Simpler models also tend to be quicker to train, while for large deep learning models specialized hardware such as graphics processing units may be important to obtain decent speed during model training. An important consideration may be to select methods that give interpretable decisions or attach a measure of certainty to their outputs, for deciding on subsequent actions. An exciting development over the last few years is that powerful and user-friendly implementations have become available, such as Weka (https://www.cs.waikato.ac.nz/ml/weka/) and Orange (https://orange.biolab.si/) which provide a graphical user interface, scikit-learn (https://scikit-learn.org/) and Tensor Flow (https://www.tensorflow.org/overview/) in python, and caret in R (https://cran.r-project.org/web/packages/caret/vignettes/caret.html). Various cloud-based machine learning platforms exist as well, including Google Cloud, Microsoft Azure, and Amazon Web Services. In the past, machine learning novices would struggle to get their first models trained and tested. However, also because of easy-to-use programming environments such as Jupyter notebook (https://jupyter.org/) or Rstudio (https://rstudio.com/), one can now instead focus on understanding how the underlying algorithms work and critically analyze the resulting models (van Dijk et al., 2021).

Machine Learning for Macroscopic Phenotypes

Collecting phenotypic data on the macroscopic scale is currently mainly a manual process, involving human experts to measure different phenotypes, which severely limits the quantity and the quality of available data. This phenotyping bottleneck slows down the understanding of the genotype-to-phenotype relations. In order to accelerate plant science, digital plant phenotyping has become an active research field with the aim to automatically derive phenotypes from sensor data (Roitsch et al., 2019). Breeders are interested in measuring phenotypes at the levels of plant organs and reproductive parts (development), the whole plant (growth), and the field (production). Leaf area, internode length, root volume, fruit size, chlorophyll content, photosynthetic activity, plant height, biomass, plant stress, water use efficiency, and yield estimation are examples of such traits. Typically, plant traits are associated with environmental parameters, such as temperature, light intensity, humidity, soil composition, and concentrations of CO_2 and O_2. Plant traits can be tracked over time to study growth and other phenological aspects. Furthermore, effects at the field level can be studied by relating the traits of neighboring plants. For the environmental parameters and some of the plant traits, sensors exist that directly measure the quantities of interest, for instance, weight, temperature, water intake, light, humidity, and gas concentrations. The data processing needed for these sensors is very limited, mainly involving noise reduction. However, as these sensors measure only at a single point in space, they cannot establish morphological and geometrical features, which are highly important for plant phenotyping. As imaging sensors are predominantly used to automatically retrieve such features, it is important to focuses on image-based plant phenotyping.

The different macroscopic levels can be studied using different imaging systems. Development at the level of plant organs can be studied in detail using microscope setups, growth of the full plant can be phenotyped in growth chambers (van Es et al., 2019) or using robotic devices in greenhouses and open fields, and drones can be used to inspect plots or complete fields (Araus et al., 2018). Different imaging sensors exist, measuring different parts of the electromagnetic spectrum, such as red-green-blue color, multi-spectral and hyper-spectral, long-wave infrared (thermal), chlorophyll-fluorescence, and tomography (magnetic resonance imaging (MRI), positron emission tomography (PET), computerized tomography (CT). While sensors and acquisition systems are now a days available, the major challenge lies in translating the high-dimensional raw imaging data into the quantification of relevant plant traits. In the past, manually engineered image-processing methods have been developed, with some success. For more complex morphological traits, however, the limits of these handcrafted methods have been reached. This is especially true when studying complex plants, in the presence of noise, and in non-controlled and cluttered environments. To deal with these complexities, the use of ML in plant phenotyping was advocated (Singh et al., 2016), in particular for image data. Classical approaches in computer vision take two steps, feature extraction using handcrafted image processing and decision-making using ML methods. Typically, supervised ML methods are used to learn from examples, which can unravel patterns in the often high-dimensional feature space that humans cannot find. The downside is that performance is limited to the quality of the handcrafted features.

In recent years, DL (deep learning) methods have become available that tackle this limitation by introducing end-to-end learning (integrating feature learning and decision-making in one frame-work) with astonishing results for general applications and agricultural applications in particular (Kamilaris and Prenafeta-Boldu, 2018).

Machine Learning for Genomic Prediction

A central objective in plant science, as well as breeding, is to explain a complex trait such as yield as a function of the available genomic, phenotypic, and environmental data. To this end, ML and other approaches aim to:

- Identify QTLs, i.e., genomic regions associated with a certain trait. This is known as QTL mapping (for experimental populations) or genome-wide association mapping (for diversity panels).
- Assess the genetic architecture: estimate the effects of individual loci and the proportion of trait variance explained by all loci combined. Related to this, genetic correlations among traits are of interest; these quantify the degree of overlap among genetic signals.
- Predict the expected trait values for new genotypes, for which only marker data are available.

Plant breeders increasingly rely on genomic selection (Crossa et al., 2017), selecting material using GPs (genomic predictions) rather than phenotypic values, and marker-assisted selection, where breeders want to obtain favorable alleles at specific loci, requiring QTL mapping. Biologists are primarily interested in genes underlying QTLs and genetic architecture. Plants can be observed at different levels (development, growth, production) using different types of sensors and sensor systems. Machine learning plays an important role in processing the sensor data to measure traits at the various levels.

Despite several potential applications of ML in linking genomes with the phenotypes, there are still several complex issues to be tackled and hence still generally, plant breeding in the private sector relies much on extensive yield evaluation across many environments, and the use of secondary traits in marker-assisted selection or genomic selection is still limited (Araus et al., 2018).

A number of efforts, from local to international, are ongoing to construct phenotyping centers, which automate and standardize high-throughput measurements of plant phenotypes at all levels – so-called phenomics (Yang et al., 2020). These will deliver more data, more objectively and accurately, yet it will take some time until sufficiently large and rich data sets are available. Developments in areas of ML that focus on the learning process itself can also help to partially circumvent this problem: synthetic data generation, through model-based simulation the measurement setup; semi-supervised learning, where only part of the data needs to be labeled; active learning, where the ML method proposes which sample should be labeled or measured next to best improve performance; and transfer learning, where a method developed for one problem is repurposed for another (van Dijk et al. 2021).

ML may prove useful to model genotype-environment interactions and to break down complex traits, such as growth or yield, into more easily measurable components, so-called secondary traits.

A future opportunity for ML is to support decision-making in a range of areas in plant research, from predicting which parts of the genome should be edited to achieve a desired phenotype (in genetic modification-GM approaches) to ensuring optimal local growth conditions by measuring crop performance in vivo in the greenhouse or on the field. While these are primarily engineering challenges, successful ML methods will offer powerful tools to researchers, particularly when they become better equipped to allow interpretation of their decisions. In this way, ML can help to address the challenges agricultural scientists face to ensure food security for growing populations in rapidly changing environments.

Major Field Crop Seed Companies Having R&D, Seed Production and Marketing in India

There are around 100 private seed companies which are selling field crop seeds in India. Based on three recognized activities of a seed company, namely, R&D, production of proprietary seeds, and sale and marketing of seeds in their own brand, about 20 % of the companies have all the three activities and intensively carry out research and produce proprietary seeds. Around 60

% of the companies are only engaged in production and marketing of seeds and these do not do any in-house research. These companies either produce OPs of public or private-bred or licensing hybrids or do in-house production. Remaining 20% are either importing seeds from other companies or sourcing produced material from the market and packing and selling the seed (Kapur, 2016). A brief introduction to few large field crop seed companies doing R & D, seed production and seed marketing is as follows.

1. Bayer

The Bayer Group comprises 392 consolidated companies in 87 countries. Global headquarters are in Leverkusen, Germany. On December 31, 2019, Bayer employed 103,824 people worldwide. Bayer is a Life Science company with a more than 150-year history and core competencies in the areas of health care and agriculture. There are three major divisions within Bayer group. These are Pharmaceuticals, Consumers Health and Crop Science. Crop Science includes following businesses.

- **Crop Protection**: Insecticides, Fungicides, Herbicides, Seed Treatment
- **Seeds**: Vegetable Seeds, Seeds for Agricultural Crops, Traits
- **Digital Farming:** Climate Field View Software
- **Environmental Science:** Professional Products

Bayer Group acquired Monsanto in June 2018, touted to be the biggest deal in history of agricultural companies, valued at USD 63 billion. With this merger, Bayer has become the world's largest conglomerate of agrochemicals and seeds. The agrochemical and seed divisions are jointly represented in Bayer Crop Science. On 13 September, 2019, Bayer closed Monsanto India Ltd integration into Bayer Crop Science. As a result of this merger, Bayer Crop Science-India divested Nunhems (Vegetable Seeds Company) to BASF and retained Seminis (Vegetable Seed Company of Monsanto) which is a world leader in vegetable breeding and innovative product development. After Monsanto merger, Bayer Group has an annual turnover of about 45 billion USD. Bayer focuses on delivering high-quality seeds, as well as cutting-edge traits for high-yielding crops with improved weed management and superior insect control.

Bayer Crop Science India has its Corporate Office at Bombay and an ultra modern mega breeding station at Bangalore to operate R & D of corn and vegetables. This mega breeding station is supported by double haploid, markers development and validation, genomics and gene discovery labs. The second major breeding station is at Hyderabad for R & D on rice, millets and cotton. A maize breeding satellite research station is at Udaipur and similarly for mustard, the satellite research station is at Faridabad. For more information, go to www.bayer.com.

On September 13, 2019, the National Company Law Tribunal approved the merger of Monsanto India Limited into Bayer Crop Science Limited. Post the merger, Monsanto products will retain their brand names and become a part of Bayer's product portfolio..This integration brings together two highly complementary businesses, creating an innovation engine for Indian agriculture. Farmers in India will benefit from Bayer's innovative crop protection products and Monsanto's expertise in seeds and traits and digital farming applications. Bayer Crop Science is a listed seed company in Bombay stock exchange. Bayer Crop Science has sound financial performance based on strong R&D and infrastructure and a motivated force of employees committed to perform (www.cropscience.bayer.in).

2. Mahyco

Mahyco, headquartered at Dawalwadi near Jalna in Maharashtra with corporate office in Mumbai has been playing a leadership role in delivering innovative crops seed to the Indian farmers including major vegetable crops, such as, chili, tomato, bottle gourd, brinjal and okra. The driving force behind Mahyco, has been its Founder-Chairman, Padma Bhushan Dr. B. R. Barwale, (1931-2017) who was a visionary and a philanthropist. World Food Prize Laureate (Late) Dr. B. R. Barwale, who is widely regarded as the father of the Indian Seed Industry founded India's first private sector Hybrid Seeds Company (MAHYCO) in 1964. During the five decades of stellar work he contributed to the food security of the country, and worked for the betterment of life of the farming community.

Mahyco Private Limited is focused on research and development, production, processing, and marketing of seeds for India's farming fraternity. Mahyco is the pioneer of high quality hybrid and open pollinated seeds. Through the use of cutting edge technology and intensive research activities, Mahyco has revolutionized the agrarian face of the country.

Despite its growth and expansion, the company remains unchanged at its core. It is still driven by a visionary zeal and the belief that it is possible to improve the lives of farmers by using science to develop superior quality agricultural products.

Mahyco's R & D amongst field crops, focuses on pearl millet hybrids, sorghum hybrids, maize hybrids, cotton hybrids, forage hybrids, mustard hybrids, castor hybrids and black mustard. For details on product portfolio, visit https://mahyco.com.

3. Rasi Seeds

Dr. M Ramasami, the founder from Agricultural family started Rasi Seeds Group in 1973 in Attur located in Salem district, Tamil Nadu, India. After establishing an unassailable brand of 'Rasi' in Cotton seed, he formed a highly experienced vegetable seeds team under the leadership of Dr. Arvind Kapur in 2009 with Corporate Office at Coimbatore, Head Office at Gurgaon and temperate vegetable breeding station at Kullu, Himachal Pradesh.

Rasi Seeds is a leading Indian agriculture company that specializes in producing quality hybrid seeds. Rasi seeds have improved the lives of many million farmers, over four decades of operations. Rasi seeds possess production farms across the country to cater the demand of Indian farmers and to address the nutritional needs of the growing Indian population.

The field crops under Rasi Seeds are cotton, paddy, maize, wheat, mustard, millets and forage crops with greater focus on hybrid seeds in cotton, paddy, maize and millets.

4. Syngenta

Syngenta India Limited, a holding company of Syngenta Global, headquartered at Switzerland, headquartered in Pune, and now acquired by Chemchina (Chinese Company) is dedicated to contributing to enrichment of farmers lives..Through world-class science, global reach and commitment to customers, Syngenta helps increase crop productivity, protect the environment and improve health and quality of life..

Syngenta includes field crops, vegetable crops and crop protection agro-chemicals in its portfolio. Syngenta seeds improve yield by prompting early emergence, vigorous growth and highest quality output. It has an established presence globally, with research and development centers, well equipped laboratories for seeds testing and efficient all - India network of field force and distributors.

The seed production is carried out through contract farmers, under stringent supervision of highly competent experts. The seed conditioning plants setup at three different locations in India cater to conditioning and packaging the seeds, protecting their essential properties and vigor.

Syngenta India continues to invest strongly in technology and marketing of field crops and vegetable seeds. State-of-the-art molecular markers technologies help broaden and deepen the germplasm base from conventional breeding. The crop range includes corn, rice, tomatoes, hot pepper, cucumbers, cabbage, cauliflower, sweet corns, beans and watermelons.

5. Nuziveedu Seed

Nuziveedu Seeds Limited (NSL), India's one of the largest hybrid seed producing companies has Registered Office at Village and Mandal, Medchal, Malkajgiri, District, Telangana and NSL Group Corporate Office at Banjara Hills, Hyderabad, Telangana. Nuziveedu Seeds, which was originally established as a part of NSL Group, has served Indian farmers for more than four decades. The company has its presence in 20 states and markets approximately 150 varieties of field and vegetable crops to large number of farmers across the country.

The genesis of the NSL Group was laid by the visionary entrepreneur, Sri Mandava Venkatramaiah, who in 1971 foresaw the opportunity of the agriculture sector and the value of hybrid cotton seeds in the Indian cotton producing market. Being inspired by his deep conviction in the cotton seed technology and his vision to make the hybrid cotton seed available to cotton growers, Sri Venkataramaiah started Nuziveedu Seeds Limited in 1973 as a proprietary enterprise which was expanded and led by his son Mr. M Prabhakar Rao who took over the reins of NSL in 1982. Under his dynamic leadership, Nuziveedu Seeds became one of the largest hybrid seed companies in India. The company has all major field crops and vegetable crops in its portfolio. NSL has developed a strong R&D programme for breeding cereals (paddy, corn, sorghum, and pearl millet), sunflower and vegetables in the past two decades. NSL has developed a strong supply chain and decentralized processing, warehousing and logistics network through its 11 processing plants, 35000 MT of conditioned storage.and nearly ~80000 sq ft of ambient storage for supply of its seed products to a wide network of 2500 distributors and 50000 + retailers. Nuziveedu Seeds acquired Pravardhan Seeds Private Limited, Yaaganti Seeds Private Limited and Prabhat Agri Biotech Limited, in 2010, 2010 and 2011 respectively.

6. Advanta Seeds

Advanta Seeds is a UPL Group company and UPL is headquartered at Dubai. Advanta Seeds - India is headquartered at Hyderabad. The Advanta Seeds South Asia region is directed from India with headquarters in Hyderabad, serving the countries, India, Pakistan, Sri Lanka, Bangladesh, Nepal and Bhutan. Advanta Seeds is a leader in vegetables and forage segment in this region, and known for high nutritional forage crops throughout the year. In Bangladesh, Advanta is the most popular brand in the corn segment and is the first to introduce forages and sunflower there. In Pakistan, Advanta's sunflower is most popular brand among the farmers. In Sri Lanka, Advanta's focus crop is corn. Among forage segment, Advanta is pioneer to introduce Sugargraze. Advanta Seeds has a significant presence in the vegetable sector with a diversified portfolio and is introducing value added products every year. In India, Advanta Seeds acquired Golden Seeds which is a flagship brand in vegetables. Advanta Seeds portfolio includes field crops, forage crops and vegetable crops.

7. JK Agri-Genetics

JK Agri Genetics Ltd. (JKAL) is a leading seed company established in 1989 with its headquarters at Hyderabad, Telangana (India). JKAL is one of the pioneers in the Indian seed industry committed to serve farming community. JKAL is engaged in research and development, production, processing and marketing of cotton, maize, paddy, pearl millet, sorghum, sunflower, castor, mustard, wheat, and sorghum Sudan grass, fodder beet, tomato, okra, chillies and other vegetable seed (www.jkagri.com).

Major Milestones

1989: Established JK Agri Genetics, (a division of JK Industries Ltd) with a view to develop high yielding varieties in various field crops, to boost agriculture productivity.

1990: Started R and D activities in sorghum, pearl millet, and corn.

1996: Started crop breeding activities in cotton.

1999: Established state of the art Biotech Research Laboratory and full-fledged biotech R and D operations started and entered into collaborative research agreement with BREF BIOTECH, IIT Kharagpur for developing. transgenic cotton hybrids.

2000: Launched high yielding rice hybrid JKRH 401.

2001: Launched DNA mapping quality assurance process for parental seed.

2003: JK Agri Genetics became an independent company, "JK Agri Genetics Ltd" and listed in the Bombay Stock Exchange (BSE).

2004: SAP package launched so as to get timely feedback on production activities, introduced "BAR Coding" in cotton seed packing to eliminate human error.

2009: Entered into licensing agreement with MAHYCO Monsanto Biotech Ltd for production and marketing of BG-II JK cotton hybrids.

2010: Established state of the art seed conditioning plant with latest world class equipment with seasonal processing capacity of 15,000 MT.

8. Corteva Agriscience

1997: DuPont becomes 100 percent owner of Pioneer.

2011: Dow and DuPont announce a definitive agreement under which the companies will merge, then subsequently spin off into three independent companies.

2015: Corteva Agriscience, agriculture division of Dow DuPont™, brand unveiled.

2019: Corteva Agriscience is founded on the rich heritages of Dow, DuPont and Pioneer. Corteva Agriscience spins from Dow DuPont, becoming a standalone company on June 1, 2019. It focuses on seed and crop protection. The newly formed corporation is headquartered in Wilmington, Delaware, but maintains global business centers in Johnston, Iowa, and Indianapolis.

The crop portfolio of Corteva Agriscience includes corn, rice, cotton, millets, soybean, mustard and vegetables. R and D centre of Corteva in India is located at Hyderabad. Corteva Agriscience inaugurated its new office premises in Hyderabad in 2019. The state-of-the-art global services centre accommodates its current 650 employees who provide business process services in the areas of finance, accounting, procurement, IT, human resources and regulatory affairs to all internal business functions in over 100 countries. The facility in Hyderabad is the global hub in a network of services centres, which also includes four regional centres of excellence in the Asturias, Shanghai, Mexico City and Wilmington.

9. Limagrain

Limagrain is an international agricultural co-operative group, specialized in field crop seeds, vegetable seeds and cereal products. Founded owned and managed by French farmers cooperative, Limagrain is the one of the largest seed companies in the world through its holding Vilmorin and Cie, European leader for functional flours through Limagrain Céréales Ingrédients, 2nd largest French baker and 3rd largest French pastry maker through Jacquet-Brossard.

The Group headquartered at Saint-Beauzire, Puy-de-Dome, France makes annual sales of 2-3 billion Euro and has a headcount of more than 10,000, spread out over 56 countries and 80 nationalities, including more than 2,000 researchers. In Auvergne the Co-operative has close to 2,000 farmer members. It conducts its business within the framework of a global, sustainable vision of agriculture and agrifood based on innovation and regulation of agricultural markets. It has 40000 ha of production under contract. In 2019, it ranked as the world's fourth largest seed producer. It invests about 15% of professional sales into research.

Corn and wheat are the two main species produced by Limagrain worldwide. Alongside this, the company develops a portfolio of species adapted to each region and market in which it operates. Due to their importance in the European domestic market, sunflower and rapeseed are strategic species. Regional species have also been identified for their essential complementarities to meet the requirements of different markets. Limagrain has introduced a plant breeding programme for species like barley, soybean, millet, rice and pulses. For others, including forage, beans and sorghum, Limagrain has distribution agreements with the breeders of these species that enable it to offer local customers a comprehensive range. The Limagrain field seeds business has traditionally been strong in Europe and North America, but since 2010 has also been expanding in new areas such as Asia, South America and Africa.

Limagrain headquartered at Hyderabad in India focuses on corn, paddy, and millets aggressively and to a limited scale on sunflower and large number of vegetables.

The Limagrain vegetable range under H. M. Clause brand is highly diverse, featuring around fifty species, including the world's most consumed vegetables, such as tomatoes, sweet and chilli peppers, onions, watermelons, carrots, cucumbers, melons, lettuce, cabbages, cauliflower, sweet corn, green beans and courgettes, as well as local species such as chicory in France, fennel in Italy, kabocha squash, mustard greens and Cantonse cabbage for Asian countries and tomatillo in Mexico. Limagrain is world leader for some of these species, such as tomatoes (the world's highest-value vegetable species), carrots, melons, cauliflower and courgette and is no. 2 in beans and no. 3 in peppers (www.limagrain.com).

10. Kaveri Seeds

Set up in 1976 by Mr. G. V. Bhaskar Rao at Hyderabad, Kaveri seeds today is India's one of the largest agriculture companies specializing in hybrid seeds in key Indian crops. Company's pursuit of excellence backed by strong Research and Development has resulted in high – yielding seeds that have made the company the trusted partner for farmers. The Company has steadily transformed products and processes to deliver the best of science, while enhancing farm productivity and ensuring food and nutritional security for millions of people. With more than 1,00,000 production growers on 26,000 hectares of land across 12 different agro-climatic zones, company's diverse portfolio of seeds (maize, paddy, cotton, sunflower, mustard, sorghum, millets, pulses, wheat, tomato, okra, chillies, watermelon, gourds, brinjal) caters to key crop segments to enable crops for diverse agro – climate and soil conditions.

Company has several satellite breeding stations (240 ha) in different agro-climatic zones of the country. The company earned more than INR 880 CR from operations and PAT of INR more than 250 CR in FY 2019-20 with employee strength of more than 1000. Company is strong in terms of financial performance, infrastructure and good sales of innovative products in large number of field and vegetable crops.

11. Ajeet Seeds

Ajeet Seeds Private Limited (ASPL), the flagship of the "Padmakar Mulay Group of Companies" headquartered at Aurangabad, Maharashtra made a humble beginning on May 13, 1986 by bringing into the market premium quality seeds of cotton, pearl millet, wheat and sorghum. Over the years the company has grown to encompass the complete spectrum of vegetables, corn and rice. Cutting edge technologies in the areas of genetics, plant breeding, biotechnology, seeds processing and marketing help company to achieve the objectives of bumper harvests, enhanced biotic and abiotic stress tolerance, improved nutritional qualities and resources conservation.

The focus is on development of hybrid seeds designed to meet the current and future national and international market demands. In the scenario where global climatic changes are an acknowledged fact, the company is poised to further strengthen developing crops which can cope with biotic and abiotic stresses. The company focuses on to take research and developments to logical conclusions, namely, development of premium quality seed, their production and finally distribution through an extensive and well established marketing network. ASPL has over 400 hectares of research farms in and around Aurangabad, 8 out-station locations and 35 multi-location testing centers spread across the country. The containment facilities for transgenic and off season crop cultivation put research efforts on a fast track.

ASPL holds in its repository diverse germplasm of field and vegetable crops. This facilitates the research scientists to embark upon genetic enhancement and crop improvement programmes. Over the years, the spectrum of seeds production has expanded to okra, eggplant, hot pepper, rice, maize, sweet corn, tomato, cucumber, watermelons, onion and gourds. The company has entered into research collaborations with international research organizations, namely, ICRISAT, IARI, IIHR, CIMMYT, AVRDC, Kasetsart University, Thailand, IRRI, Philippines, to further develop and produce field and vegetable crops. ASPL is in a position to serve cross cultural demands and a plethora of cropping patterns in India. Company's market presence has grown appreciably and the activities of Ajeet Seeds Pvt. Ltd. have extended and spilled over to its subsidiary seed company, "Arya Hybrid Seeds Ltd".

ASPL has commissioned modern seed conditioning plants at Chitegaon and Phulambri in Aurangabad, Maharashtra. The Company is now poised to launch its products in the ASEAN countries.

2

Seed Business

Indian Council of Food and Agriculture organized SEED WORLD-2019 (World Seed Trade and Technology Congress) from 18-21 September, 2019 at The Hotel Lalit Ashok, Bangalore and presented SEED INDUSTRY SCENARIO REPORT (ICFA, 2019; www.icfa.org.inwww.seedworld.in). The information as given below is largely based on this report and other presentations made during the congress, particularly one by Kapur (2019).

Global Seed Market

The global seeds market was valued at USD 59.71 billion in 2018, exhibiting a CAGR (compound annual growth rate) of 7% during the year 2011 - 2018. There are expectations that the global seed market may reach to USD 90.37 billion in 2024 witnessing a CAGR of 7.9% during the forecast period 2019-2024. The growing demand for grains, oils, and vegetables is a significant driver for expanding the seeds market. Grains represent the biggest portion of daily calorie intake in developing countries of the Middle East and Africa, Asia/Oceania, and CIS, and therefore, the demand for grains will be one of the most critical stimulants for the expansion of the seed market. The shift in farming practices worldwide has necessitated adoption of commercially produced enhanced seed varieties by the farmers as opposed to using seeds from the last harvest. High yield, improved nutritional quality, reduced crop damage, disease resistance, etc. motivated farmers in investing in commercial seeds which alone account for about 50% of the gain in productivity.

The massive increase in the demand for bio-fuel, buoyed by the large subsidies provided in many western countries, has increased production of global bio-fuels by almost 150% between 2004 and 2010 from 42 billion liters to 104 billion liters respectively. As per FAO report, the bio-fuel boom had a major impact on the evolution of world food demand for cereals and vegetable oils. Moreover, it states that without bio-fuel, the growth rate of world cereal consumption is equal to 1.3%, as compared to 1.8% of bio-fuel and thus is an important factor behind expansion of global seed market. Besides these, acceptance of area under genetically modified crops, increasing demand for animal feed, rapid adoption of biotech crops and decreasing per capita farm land are contributing to the growth of global seed market. On the other hand factors like rising concerns over GM seeds, years involved in development of new traits and long GM approval timelines and government regulation have been holding the sector back. The North American seeds market has been valued at USD 20.91 billion in 2018, serving the farmers mainly in grains, cereals, fruits, vegetables, and oil and forage crops. The North American seed market is expected to reach USD 30.9 billion by 2024 and is estimated to register a CAGR of 6.46%, during the forecast period. North America is the largest commercial seeds market, accounting for more than 35% of the global seed market share closely followed by China.

According to Kapur (2019), the current total global seed market is estimated to as 60 billion USD out of which traded seed market accounts for 45 billion USD. The value of seed market of 12 major countries is given in Table 2.1 and also presented in Fig. 2.1 (Kapur, 2019). The global seed market divided into GM vs. conventional seed from 2013 to 2018 is given in Table 2.2 (McDougall, 2019). The global seed market split is given in Table 2.3 (Kapur, 2019). From Table 2.1 and Fig. 2.1, it is obvious that USA leads the global seed market closely followed by China. India stands at fourth place with domestic seed market size of 3.8 billion USD (=Rs. 28700 CR) as in 2018.

Table 2.1: Domestic seed market in value term (billion USD) for top 12 countries in the world

Country	Estimated value of domestic market (billion USD)
USA	12.00
China	10.00
France	3.90
India	3.80
Brazil	2.70
Japan	1.90
Germany	1.70
Italy	1.00
Argentina	0.90
Canada	0.80
Russian Federation	0.70
Spain	0.60
Total	40.0

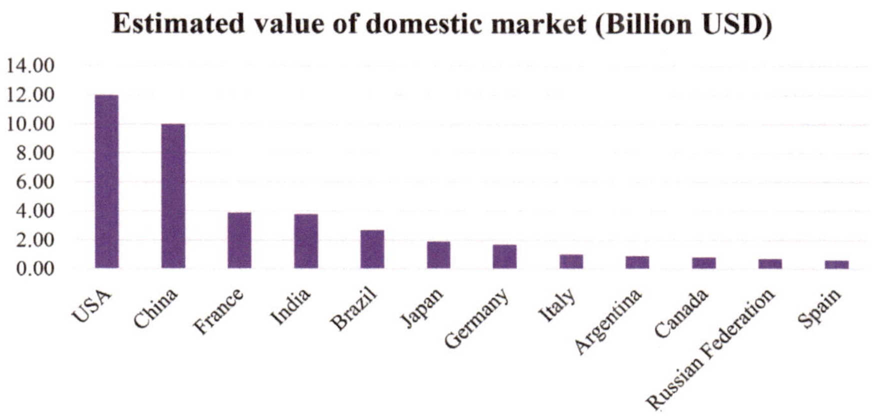

Fig. 2.1: Domestic seed market of top 12 countries in the world

Table 2.2: Commercial global seed market (2013-2018, billion USD)

	2013	2014	2015	2016	2017	2018
GM seed	20.100	21.054	19.789	20.039	22.206	21.970
Conventional seed	19.282	19.481	17.441	16.846	18.912	19.700
World seed market	39.382	40.535	37.230	36.885	41.118	41.670

Table 2.2 clearly shows that proportion of GM seed and conventional/non-GM seed market in the world from 2013 to 2018 has remained almost equal. It should be kept in mind that the major contributors to the GM seed market are USA, Argentina, Brazil, Canada, India and China. The figures quoted from various sources are not exactly tallying and there are minor differences and this is due to the fact that the quoted sources have used varying information emanating from different sources and there is some degree of estimation involved based on market intelligence. But the differences are minor. It is fair to reiterate that the total market value of seed produced globally in 2018 has been 60 billion USD and the value of traded seed globally is close to 45 billion USD.

Table 2.3: Global seed market split in term of major seed crops

Crop	Seed market (%)
Corn	32
Vegetables	18
Soybean	12
Rice	10
Canola	5
Wheat	4
Potato	4
Cotton	3
Barley	2
Sunflower	2
Sugar beet	2
Others	6
Total	100

Globally, corn seed is the market leader followed by vegetables, soybean and rice. As a matter of fact corn, vegetables, soybean and rice together account for 72 % of the total global seed market as in 2018 (Kapur, 2019).

Indian Seed Market

The profile of seeds in India has changed over the years. Earlier, it was the seeds saved from the previous crop that was used in Indian agriculture. Now the trend is changing and it is the most advanced seeds developed by seed companies and a few ICAR crop based institutes and state agricultural universities that dominate the farmers' fields. The phenomenon has roots in the changing dynamics of agriculture, not only in India but also worldwide. The spectra of changing climate, the danger of depleting resources and the threat of burgeoning population has diminished the productivity of agriculture. Stagnant yields and yield loss have become quite persistent. Adding to the chaos, India has its own share of problems –lower penetration of technology, shrinking land holdings, marginal farmers, lack of mechanization, shortage of labours to name among a few.

In 2018, the Indian seeds market reached a value of US$ 4.1 Billion, registering a CAGR of 15.7% during 2011-2018. It is further expected to grow at a CAGR of 13.6% during 2019-2024, reaching a value of US$ 9.1 Billion by 2024. Coupled with increasing domestic demand and demand for quality seeds in various foreign countries, mainly the South East Asian countries, seed industry in India is witnessing new paradigms of growth and development. The use of hybrid seeds has silently but consistently witnessed growth along with several other driving factors like increasing middle class and increasing disposable income, growth in the food processing sector, increasing seed replacement ratio and other allied factors. Rising awareness among the farmers related to the benefits of using quality seeds has led to an increase in the demand for seeds over the past few years. This has resulted in an increasing willingness among the farmers to pay higher price for quality seeds provided they get value out of the new seed. When compared to the global seed production, India's share is very less. India is way behind countries like USA and China in terms of total seed market size. US seed market is over 12 billion USD, that of China is over 10 billion USD and Indian seed market size is worth 4 billion USD as in 2018.

Of the global seeds market, India’s market share was 4% in 2017 which was dominated by cotton seeds, maize seeds and fruits and vegetable seeds. India stands 16th globally in fruits and vegetables seed exports, and had nearly 2% share of the global trade in 2016. In terms of global trade, India is almost self-sufficient in flower, fruits and vegetables and field crops seeds. The Indian seed market is majorly dominated by non-vegetable seeds such as corn, cotton, paddy, wheat, sorghum, sunflower and millets. Grain seeds represent the largest seed type, accounting for more than half of the total seed production. Uttar Pradesh represents the largest producer, accounting for around 12% of the total market share of grain seed production. Cotton is one of the most cultivated crops in India, with great economic importance attached to it. The cotton seed market was valued at Rs. 24.86 billion (Rs. 2486 CR) in India in 2016. India is also the second largest producer of cotton worldwide. Since 2002, Bt. cotton has steadily prevailed over India’s cotton fields raising overall cotton production to record highs. Today, over 90% of cotton grown in India is cultivated by using Bt. cotton seeds. No crop could match this kind of seed replacement rate as witnessed in case of Bt. cotton in India despite the fact that Bt cotton seed is costlier than the non-GM/conventional cotton seed. Bt. cotton seed market in India is about 4.5 CR packets of cotton seed each of 450 g Bt. seed. There are very few countries in Asia and Africa which allow commercial cultivation of GM crops and hence cotton seed export has not picked up in last decade. Now with more countries opening up for GM cultivation, there is huge potential to export cotton seeds from India. Maize is one of the largest consumed commercial seed, accounting for Rs.14.91 billion (Rs. 1491 CR) of revenue in 2016. The growth in maize is attributed to its increased use as livestock feed and in ethanol production. Presently, approximately 25% of maize is used as food grain, while remaining 75% is used to meet nonfood demand, viz. bio-fuels, poultry feed, animal feed, brewing alcohol, starch based wet milling industries and other industrial uses. However, the enhanced investment in maize by global seed companies would flow into India through technology transfer and Public-Private Partnerships in the next fifty years. Government of India has not yet fixed any target for increasing acreage of hybrid rice in the country. However, efforts are being made to promote cultivation of hybrid rice through various crop development programmes such as National Food Security Mission (NFSM), Bringing Green Revolution to Eastern India (BGREI) and Rashtriya Krishi Vikas Yojana (RKVY). With private sector playing a major role in hybridization of rice, the hybrid rice seed market has touched more than 45 thousand metric tons volume sale in 2016. The growth in 2016 can be attributed to significant increase under Kharif acreage of rice due to abundant rainfall, after a consecutive two years of drought and growers shift from inbreds/OPs to hybrid due to problems faced previously with OPV's.

Being agriculture based country; India is probably best placed to cater to not only domestic but also global seed requirements. Its importance is also reflected by its overall position in the global seed trade. In case of fruits and vegetables, India ranks number 16 in the global fruits and vegetable seed exports with only 1.68% share of the global trade in 2014-15. Although, there is a substantial increase in India's export of fruits and vegetable seeds in value, India has already started to lose a large share of such exports over the past few years. Lack of appropriate policy reform in Indian seed sector can be one missing element, which are discouraging exporters and producers to engage more in export of fruits and vegetable seeds. It also needs to be examined whether the bargaining positions of other countries have improved or they have taken any competitive advantage or market imperfections of India has increased in recent years. However, India's imports of the fruits and vegetable seeds have depicted a declining trend in 2016, when compared to 2015. This decline is reflected not only in quantity but also in value terms. The major countries exporting fruits and vegetable seeds to India in 2016 included Thailand, Egypt, Chile, United States and China, accounting 64% of the total imports of fruits and vegetable seeds, amounting to Rs.4.16 billion (INR 416 CR) in India. The Indian seeds market is anticipated to grow at a considerable CAGR rate due to improvement of seed replacement rate, production and distribution of quality seeds appropriate to agro-climatic zone at affordable prices along with a determined effort to address region specific constraints. Moreover, several factors, including increased subsidies and renewed government thrust on the use of high yielding varieties, will lead to an increased productivity in the seed market. The total export of fruit and vegetable seeds in 2018-19 was valued at Rs. 84927 lakhs (INR 849 CR). The Indian seed market has witnessed a major restructuring as a result of the implementation of some progressive policies by the government. Seed Development, 1988 and National Seed Policy, 2002 have helped in strengthening the Indian seed industry in the areas of R and D, product development, supply chain management and quality assurance. Owing to this, India has emerged as the fifth largest seed market across the globe. Moreover, the active participation of both, public and private sectors has also played a vital role in laying a strong foundation of the seed industry.

The value of India seed market for the years 2010-18 is shown in Table 2.4 and Fig. 2.2 (ICFA, 2019). On an average, current value of Indian seed market is worth Rs. 30000 CR out of which vegetable seeds alone account for Rs. 4000 CR in 2019.

Table 2.4: Indian seed market in value term for 2010 to 2018

Year	Value (billion INR)	INR in CR
2010	46	4600
2011	59	5900
2012	86	8600
2013	103	10300
2014	122	12200
2015	141	14100
2016	166	16600
2017	234	23400
2018	287	28700

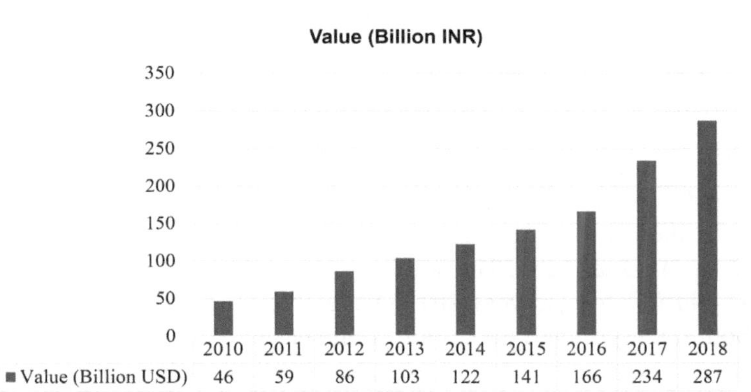

Fig. 2.2: Indian seed market in value term from 2010 to 2018

Indian seed industry has seen stupendous growth in recent times. India's need to etch a remarkable agriculture story has fuelled this growth. Conducive policy reforms and government support have spurred the transformation of this industry in recent times. Many regional and multinational players have captured the scene and today private sector is a formidable presence in India. Private sector has played a major role in changing India's seed sector. Investment and technical expertise garnered from different parts of the world has made this fete possible. The seed industry of the country is going to align with the changing food consumption scenario of the country. There has been a decline in the proportion of expenditure on food items in last three decades. The change has occurred in both urban and rural areas. Between 1990 and 2010, consumption of cereals and pulses has decreased considerably while consumption of fruits and vegetables has almost doubled. The future seed market of the country is expected to witness more adoption of hybrid seeds. The country is going to witness an increase in the cultivation of vegetables and fruits, driven by the growth and development of the food processing sector along with changing consumption habit of the people. The export sector is also set to witness growth. One of the most important developments in the future will be the increased demand for quality vegetable and fruit products, because of the increased consumer awareness and increase in disposable income of the population. As a result of this, trend of protected cultivation will increase in India. The current seed industry might also need to develop new products in order to attune with the greater mechanization of agriculture sector in India in the near future, vastly due to labor challenges. One more expected change that the seed industry is going to witness is in the form of increased demand for fruits and vegetable varieties with higher nutritional values. Life style related health problems are increasing in the country with diseases like obesity, diabetes, malnutrition etc. on the rise and hence people are becoming more health conscious which will propel the use of fruits and vegetables with more nutritional values.

International Seed Federation (www.worldseed.org) has compiled figures on seed export and import for each of the seed producing countries for several years. The latest figures are available for the year 2017. The same is reproduced in Table 2.5 for India to get an idea of seed volume and seed value for the exported and the imported seed. The seed export stands at Rs. 750 CR and the import at Rs. 900 CR for the year 2017. Vegetable seed has been the major component both for export and the import. India has potential to emerge as the strong seed importing country in future due to varies agro-climatic conditions and entrepreneurship skill of the Indian farmers.

Table 2.5: Seed export and import by India during 2017

	In volume term (tons)			
	Vegetable seed	Flower seed	Field crop seed	Total
Export	7383	97	25556	33036
Import	2789	7	26660	29456
		In value term (million USD)		
Export	66	2	33	101 (Rs. 750 CR)
Import	77	9	35	121 (Rs. 900 CR)

Main Events in the Seed Sector in Post-Independence Era

The seminal role of the quality seed is in achieving sustainable agricultural production through enhancing crop productivity. Hence smooth flow of quality seed from breeders' plots to farmers' fields is a *sine qua non* for agricultural prosperity. India attained its all time high food grains production of 285 million tons during 2018-19, from a meagre 50.5 million tons during 1950-51. This is largely attributed to the use of quality seed of improved varieties/hybrids of crops and adoption of modern production technologies by the farmers. The main events facilitating seed sector development since the early 1950s are listed as follows (Singh et al., 2015; Chauhan et al., 2020).

Year	Event
1952	An Expert Standing Committee was set up for Seed Improvement Programme; Grow More Food Committee confirmed an unsatisfactory performance of seed activity.
1955	Essential Commodities Act.
1957	Indian Council of Agricultural Research started All-India Coordinated Maize Improvement Project in collaboration with Rockefeller Foundation.
1959	First Indo American Agricultural Production Team of Ford Foundation, headed by Dr. S. E. Johnson stressed the use of improved seeds.
1959	Prof. A. S. Carter of USA furnished a Model State Seed Law.
1960	Coordinated projects on sorghum and pearl millet started.
1960	First Agricultural University (UPAU) on land grant pattern was established at Pantnagar (initially in UP and now in Uttarakhand), also hosting of the Tarai Development Corporation (TDC). Now this university is known as G.B. Pant University of Agriculture and Technology (GBPUAT), Pantnagar (Uttarakhand).
1961	First hybrid maize was released from ICAR-maize improvement project.
1961	2,500 Seed Multiplication Farms were set-up.
1961	First Seed Testing Laboratory was established at the IARI, New Delhi.
1963	National Seed Corporation was established. The NSC imported 18,000 tons of wheat- seeds from Mexico; first hybrid sorghum was released from ICAR sorghum improvement project.
1964	First Bajra (pearl millet) hybrid was released.
1965	AICRP on wheat was established.
1966	Seed Act was passed by the Parliament.
1967	Seed Testing Manual was published.

1968 Seed Rules were framed by the Government of India.
1969 Seed Act-1966 came into effect.
1969 First State Seed Corporation was established at Pantnagar (initially in UP and now in Uttarakhand).
1969 Govt. of India established State Farms Corporation of India; UP Seeds and Tarai Development Corporation established.
1970 The Patents Act (with amendments in 1999, 2002, 2005).
1971 Indian Society of Seed Technologists was formed; National Commission on Agriculture constituted.
1971 Seed Certification Standards were published.
1974 National Seed Project was launched.
1975 National Commission on Agriculture submitted project report on NSP (national seed project).
1977-78 NSP-I launched with the World Bank assistance of US$ 52.7 million.
1978-79 NSP-II launched with the World Bank assistance of US$ 34.9 million.
1979-80 All-India Coordinated "National Seed Project" launched; All-India Coordinated Project on "Seed Borne Diseases" launched.
1981 First workshop on seed technology was held under NSP.
1983 Seed Control Order.
1984 The Plants, Fruits and Seeds Order.
1986 Consumer Protection Act enacted.
1987 Seed Transport Subsidy Scheme.
1988 Separate Seed Division at ICAR created.
1988 New Policy on Seed Development.
1988 Indian Minimum Seed Certification Standards revised.
1989-90 Special Project on hybrids in nine selected crops and seed technology launched.
1990-91 NSP III launched.
1991 AICRP on Seed-borne Diseases merged with NSP (Crops).
1994 Government of India signed GATT Agreement.
1999 Geographical Indications of Goods (Registration and Protection Act).
2000 Seed Bank Scheme.
2001 PPV&FR Act-2001 passed.
2002 Seed Sector Reforms – National Seed Policy.
2004 New Seed Bill proposed, by amalgamating the provisions of Seeds Act, 1966 and Seeds Control Order 1983 as a single enactment.
2004 Establishment of ICAR-Directorate of Seed Research, Mau (UP).
2004 Introduction of Seed Bill in parliament.
2005 National Seed Plan and launching of ICAR-Mega Seed Project.
2005 WTO – SPS (Sanitary and Phytosanitary measures).
2006 ICAR Mega Seed Project on seed production in agricultural crops, horticulture and fisheries launched.
2007 National Food Security Mission.
2007 Rashtriya Krishi Vikas Yojna.
2007 National Biotechnology Development Strategy.
2007-08 Formation of Expert Committee to suggest measures for improvement of functioning of the state seed corporations by Department of Agriculture and Cooperation, Government of India.

2008	Joining of OECD Seed Schemes by India and presently 249 varieties of 20 crops have been enlisted for seed export under the scheme.
2009	National Biotechnology Regulatory Authority of India Bill.
2009-14	Launching of seed export-import policy.
2011	Modified New Policy on Seed Development.
2015	ISTA accreditation of first public sector laboratory.
2015	Up-gradation of ICAR-Directorate of Seed Research to ICAR-Indian Institute Seed Science, Mau, UP.
2016	Cotton Seed Price (Control) Order and Licensing and Formats for GM Technology Agreement Guidelines.
2019	Seed Bill-2004 with modifications under discussion as Seed Bill-2019.

Role of the Public Sector in Seeds

The public sector comprises one national level corporation, National Seed Corporation (NSC), 15 State Seed Corporations (SSCs), 22 Seed Certification Agencies (SCAs), two Central Seed Testing Laboratories and 122 State Seed Testing Laboratories (3 ISTA accredited and 20 have ISTA membership) which is providing requisite strength in serving seed industry and farmers. The research and development in the public sector is dependent on public research under the aegis of the ICAR institutes and SAUs, which are also engaged in production of breeder, foundation and certified/truthfully labeled seed of their varieties. Seed is also produced by farmers under Farmers' Participatory Programme of several Institutes and under Seed Village Programme of the Government of India. As many as more than 35 agricultural universities and 22 ICAR institutes across the country are engaged in seed production activities. The public-private-farmer research-extension-production should be effectively linked to achieve accelerated growth. The ICAR had launched the Mega Seed Project in 10th Five Year Plan from 2005-06, and it has been continuing in 12th Five Year Plan under Consortium Research Platform. Directorate of Seed Research, Mau in Uttar Pradesh has been entrusted with the responsibility of coordinating the activities under the ICAR seed project "Seed Production in Agricultural Crops".

As per recent update by Chauhan et al. (2020), currently, public seed sector comprises National Agriculture Research, Education and Extension System (NAREES) having 64 ICAR Research Institutes, 6 Bureaux, 15 National Research Centres,13 Directorates/Project Directorates, 3 Central Agricultural Universities and 5 Deemed Universities, 82 All India Coordinated and Network Projects (59 All India Coordinated, 21 Network and 2 other projects), 11 Agricultural Technology Application Research Institutes (ATARI), 716 Krishi Vigyan Kendras, 4 Central Universities with Agriculture faculty, 63 State Agricultural Universities (https://www.icar.org. July 24, 2020), National Seed Corporation (NSC), New Delhi with 10 Regional and 66 Area Offices, 16

State Seed Corporations, 25 State Seed Certification Agencies and 134 State Seed Testing Laboratories; 18 ISTA members and six (one public and five private sector) ISTA accredited laboratories. However, these high sounding numbers don't match up to the expected out-put except few exceptions.

Role of the Private Sector in Seeds

The New Seed Development Policy (1988-89) has been a significant milestone in the Indian Seed Industry, which transformed very character of the seed industry. It gave Indian farmers access to best seed and planting material available anywhere in the world. The policy stimulated appreciable investments by private individuals, Indian Corporate and Multinational Companies in the Indian seed sector with strong R&D base for product development in each of the seed companies with more emphasis on high-value hybrids of cereals and vegetables and hi-tech products such as Bt cotton. The national economic liberalization regime introduced in 1991 further accelerated growth of private seed industry in the country. The New Policy on Seed Development liberalized to a great extent import of vegetable and flower seeds under open general license (OGL) in general and seeds of other commodities in a restricted manner and also encouraged multinational seed companies to enter seed business. To begin with more than 24 companies initiated research and development activities and made substantial commitments for investment on research and development in response to this policy initiative. Now the number of private seed companies has gone beyond 300. The major players in private seed sector with strong R&D, seed production and sales and marketing in India are Bayer Crop Science including Monsanto and Seminis (acquired by Bayer in 2018), Corteva (an agricultural company-as a result of spin off from the merger of Pioneer-Dupont-Dow Agri-Science), BASF, Syngenta, Mahyco, IAHS, Sakata, Takii, Tokita, Nuziveedu, JK Agri-Genetics, Kaveri, Limagrain, VNR, Tata Rallis, Rasi, ACSEN HYVEG, Ajeet, Ankur and Advanta, etc. The investments are expected to increase with increasing volumes of seeds of proprietary hybrids and preparedness of farmers to pay higher price for quality seed.

As a result of these interventions, the Indian seed industry has evolved by adopting innovation and scientific advancements in the areas of variety development and quality seed production. From an essentially non-existent seed industry five decades ago, it is truly an industry involving millions of farmer producers, hundreds of seed companies (both public and private), ably serving the cause of Indian agriculture. The Indian seed industry is a vibrant industry, consisting of players from the public (43%) and private (57%) sectors. The public sector, in general, is dealing with the high volume, low-value crops, and private seed sector is more engaged in low volume and high-value seed crops. But with the licensing of large number of wheat, rice and maize varieties during the last 2-3 years this notion is changing. Indian seed industry is the fourth largest seed consumer in the world, with 8% of global seed market, after US, China and France. During the last five years Indian seed industry has been growing at a compound annual growth rate of 15% from Rs. 5,000 crore in 2009-10 to Rs.12200 crore in 2014 and at Rs. 30000 crore in 2018. This industry is undergoing wide ranging transformation including increased role of private sector, entry of multinational companies MNC), joint ventures of Indian companies with MNCs and consolidations. Indian seed industry is poised to grow at a rate of 12% for next five years, and India will soon be ranked at number 2 or 3 in Global Seed Business.

The Prospects

The Indian seed system is endowed with following strengths: (i) diverse agro-climates across the country suitable for production of seed of diverse crops, (ii) rich biodiversity resources (more than 4,500 HYVs in different crops available) (iii) well-trained and experienced scientific and technical work-force, (iv) strong conventional breeding as well as molecular breeding support, (v) product-directed research and generation of new technologies for management of seed system, (vi) Assured end-to-end (breeding, production, logistics and marketing chain) approach (vii) national level giant organizations like National Seed Corporation (NSC) and State Farms Corporation of India (SFCI), 25 state seed certification agencies, 16 state seed corporations, and 108 notified seed testing laboratories, (viii) state assisted farmers co-operatives for foundation and certified seed production, and (ix) fast developing private seed sector,~500 large and medium size private seed companies (around 24 have R&D establishment) recognized by the DST.

Given its high agro-ecological diversity, favourable agro-climatic zones, vast irrigated lands and rich biodiversity, India is ideally suited to be a global hub for production and distribution of quality seed of a large number of Agri-Horticultural crops. The ongoing public private sector collaboration in the seed sector can further be strengthened, streamlined, and branded. International collaborations, particularly South-South collaboration with Asian and African countries, will greatly boost seed economy as well as food security. In order to increase competitiveness of the Indian seed sector and to increase cost effectiveness, high priority should be given to seed science and research and technology development on the following lines.

- Strengthen maintenance breeding and develop diagnostics for genetic purity.
- Deploy genomics to discover gene/s governing dormancy in rice, seed germination and seed longevity in soybean and stress - tolerant genes to produce superior quality seeds.
- Use transcriptomics of seed development to unzip molecular regulation of improved seed characters.
- Identify suitable molecular markers for MAS in improving seed attributes and developing mapping populations to tag gene/s governing important seed attributes in different crops by hybridizing contrasting parents for important seed characters like dormancy, longevity, seed emergence, seed vigour etc.
- Promote nano technology to dispense nutrients through nano pores during seed germination and to avert entry of pathogens.
- Synergistically use conventional seed technology with frontier science in relation to seed dormancy in rice, longevity in soybean, low multiplication rate in groundnut, genetic purity in sunflower.
- Revisit seed certification standards under the backdrop of OECD scheme of seed certification.
- Adopt measures to ensure quality seed production in the era of climate change.
- Agro-ecologically delineate AEZs for customized production of quality seed of specific crops.
- Promote participatory seed production and entrepreneurial employment of rural-youth along the entire seed chain.
- Strengthen human resources in the seed sector and enhance farmers' skill to up-grade the quality of farm-saved seed, and augment the extension services.
- University curricula on seeds should ensure integration of seed science, seed technology, seed production, quality standards, marketing, pricing, seed business and distribution.
- Creating enabling environment for hassle free operation of seed companies with due regards to IPR regulations.

3

Commercial Plant Breeding Programme

Any commercial plant breeding programme run by the private seed companies typically called as Research and Development (R & D) is carried out through various models operating. There is no fixed organizational set up and different companies adopt the working set up as per size of their programme, availability of manpower and resources, area of the breeding station and several other considerations. One thing is clear that any commercial plant breeding programme will have more numbers of crop breeders than the supporting scientists. The core research activities and accordingly deployment of the manpower at any crop breeding R&D station are as follows.

Individual Crop Breeding

As per policy of the company, first major crops are identified and then these are prioritized to deploy the manpower and the resources. Once the crops have been identified which in itself is a strategic exercise based on discussion among the divisional heads of breeding, multi-location testing, seed production, sales and marketing and product development and finally endorsed by the top management of the company, the breeders are allotted to the crops. Depending upon the size of any crop breeding programme, there could be one breeder for one specific crop, there could be more breeders in the same crop, there could be a single breeder for a group of crops as the situation may demand. The individual crop breeding programme involves the following activities which could be handled by the same breeder or which could be handled by separate scientists depending upon the size and the strength of the crop breeding programme. The crop breeding programmes are supported by the plant pathologists and entomologists who are not allotted individual crops, instead they work as a common resource for the entire R&D.

Germplasm Management

This section involves germplasm collection including local collection, institutional collection and germplasm seed import from potential sources abroad, germplasm characterization, evaluation, germplasm enhancement, multiplication and proper seed storage. The potential international sources for most of the field crops are CGIAR funded International Agriculture Research Centres like CIMMYT for wheat and maize, IRRI for rice, ICRISAT for groundnut, chick pea, sorghum, pearl millet, ICARDA for lentil, IITA for grain cow pea and CIAT for dry beans. The selected elite germplasm lines are utilized in the regular breeding programme directly as such or as parents in the pedigree/bulk/SSD method of advancement of segregating generations or as donors in the back-cross breeding programme. This activity can be handled by a separate germplasm scientist meant exclusively for this job or by the crop breeder himself or the team of the crop breeders.

Line Breeding/Line Development

This section deals with development of new breeding lines/inbred lines to be used as parents in the development of commercial hybrids or even as OPs in case any seed company is interested in marketing of seed of open pollinated cultivar(s). In case of cross-pollinated crops, usually no seed company will go for OPs. But in case of self-pollinated crops, this option is open to the private seed companies. The lines are developed following regular plant breeding procedures like making a cross between two parents/making a cross between two F_1 hybrids/making multiple crosses as shown below.

i) A x B

ii) (A x B) x (C x D)

iii) (A x B) x C

iii) (A x B) x (C x D) x (E x F) x (G x H)

Simply stated, the first and the most commonly practiced option is a single cross, the second and the third the next preferred options are a double cross and three-way cross, the fourth and the least preferred option is the multiple cross which is a cross between two double cross hybrids. In case, more characters from diverse sources are to be combined into one line, then the F_1s of single, three-way crosses and the double crosses are preferred. The F_2 is the first segregating generation in case of the first option and the F_1 itself is the segregating generation in the second, third and the fourth options.

Two-parent crosses are the most common type of population in self-pollinated crops. However, populations developed from more than two parents have also been used in line breeding. Multiple parents are often used to transfer genes for pest resistance from agronomically unacceptable parents into more desirable populations.

Useful segregating populations have also been developed by the use of a limited number of back-crosses. One purpose of back-crossing is to increase the frequency of segregates beyond that available in a two-parent population. For yield improvement of large seeded cultivars, it is desirable to obtain the higher yield potential available in commercial cultivars with smaller seed. A cross between a large- and a small-seeded parent does not provide an adequate frequency of large-seeded progeny (Bravo et al., 1981). A back-cross to the large-seeded parent results in the necessary large-seeded segregates.

A limited number of back-crosses have been used to transfer genes from a plant introduction to a high yielding genetic background. The technique is especially useful when character being transferred is controlled by multiple genes which necessitate the screening of a large number of segregates each generation to recover those with desirable genes for use in crossing. When different recurrent parents are used during back-crossing, the procedure is sometimes referred to as modified back-crossing (Fehr, 1987). There are two reasons for changing recurrent parent during back-crossing. First, the original recurrent parent may be replaced by one with more desirable characteristics. Second, the different parents may be used to accumulate different favourable characteristics from each.

The resultant segregating generations are routed through typical pedigree method of generation advancement. However, lately bulk and SSD till F_4 generation are being preferred to facilitate rapid generation advancement under controlled greenhouses followed by application of pedigree method in strict sense in F_5 and F_6 generations. The end result of this exercise is development of superior inbred lines to be used in hybrid making.

Hybrid Making and Evaluation/Commercial Breeding

Once large numbers of superior inbreds/lines are available through line development or line breeding programme, paired crosses are usually made. While making such crosses, it is seen that the crosses between lines originating from the same source are avoided and the crosses between lines originating from different sources are encouraged. For example, the crosses among lines derived from A x B should be avoided and the lines derived from A x B should preferably be crosses with the lines derived from C x D. However, this cannot be generalized as several other factors like synchronous maturity, complementarity of the parental traits and combining ability have also to be factored in. Thus, it is necessary to screen the inbred lines for their general combining ability also through top cross or diallel cross mating and in this process quite several lines are dropped from commercial cross making. The resultant F_1 hybrids are evaluated at the R&D station following appropriate statistical designs (the most preferred being RBD) in the initial evaluation trial (first year), advance trial-I (second year) and advance trial-II (third year) along with the market leading checks. While advancing entries from IET to AVT-I to AVT-II, the poor performers are dropped from further testing. One of the common criteria while advancing the entries from one trial to the next is often at least 10% significant yield superiority over the best check/reasonably numerical yield superiority over the best check and/or add on value in terms of resistance to certain major diseases/tolerance to serious insect-pests/nutritional improvement. Nutritionally improved cultivars are now being called as bio-fortified cultivars .

Multi-Location Trials (MLT)

The top performing hybrids based on evaluation in advance trial-II, are handed over to the trial department or the scientists engaged in conduct of MLTs. This responsibility cannot be handled by the specific crop breeders as they are fully occupied with on the breeding station R&D activities almost round and further, the MLTs are conducted at multi-locations spread over regions where the hybrids may have markets. The MLTs are conducted at the regional research stations of the company and also on farmers' field by the scientists specifically deployed for this purpose. The crop breeders stationed at the R&D Research Farm visit these trials frequently particularly at flowering and maturity stage to have on the spot assessment based on various characters appealing to the eyes. The data is recorded by the MLT-team.

Exploitation of GEI in Breeding Programmes: The presence of large GEI (genotype x environment interaction) necessitates the development of varieties suited to different agro-environments based on their adaptability /stability/ characteristics and/or selecting genotypes that perform well over many sites. The first approach would yield greater genetic gains but cost would likely be higher. The second approach is less expensive but the gains would also be less.

LeClerg (1966) suggested that in early stages of testing programmes, it would be better to have as many locations as possible but with only one replication per location vis-à-vis multiple replications at a fewer locations. Kang (1993) discussed the disadvantages of discarding genotypes evaluated in only one environment in early stages of a breeding programme. Some potentially useful genes could be lost from gene pool due to limited testing. The discarded genotypes could have the potential to do well at another location or in another year. To fall in line with this arguments, ICAR institutions and state agricultural universities rely on comprehensive testing of their breeding lines/products at least for three years spread over a large number of locations throughout the country under all India coordinated research projects (AICRPs) on almost all the major field and vegetable crops. Private sector seed companies also rely on this and conduct large number of multi-location trials of their advanced products either on farmers' field or on their own research stations located in strategic areas as per specific needs of the crop. To facilitate this testing and to get better results, private seed companies identify few mega-environments suiting to a particular crop and ensure that their products are certainly tested in the mega-environments as per crop/product demands. The companies take the feed-back from sales and marketing teams also to decide specific locations in the mega-environments. Considering the importance of GE interaction, private sector seed companies enter their selected products into AICRP trials on payment basis but they have their own independent "Testing Units/ Departments" within over all umbrellas of R & D and attach serious attention to this exercise and ensure timely seed dispatch using air cargo/courier services, timely trial planting and submission of the results in time bound manner so as to arrive on quickest possible decision for promotion of elite entries to the next stage of testing/seed production/commercialization. Quite aptly, private sector seed companies' Leads/Heads of Department/Directors often say that "Seed Business" is a "Speed Business".

A large GE interaction reflects the need for testing cultivars in numerous environments (locations and/or years) to obtain reliable results. If the weather patterns and/or management practices differ in target areas, testing must be done at several sites in the target areas. If cultivars are to be used in marginal areas, testing must begin in those areas as early as possible.

Multi-environment testing makes it possible to identify cultivars that perform consistently from year to year (reflecting small temporal variability) and those that perform consistently from location to location (reflecting small spatial variability). Temporal stability is desired by and beneficial to growers but the spatial stability is desired by and beneficial to seed companies and the breeders.

An understanding of genotype responses to individual factors helps in interpreting the results and making good decision. When an environment/ location is in drought prone or flood prone area, then that environment can be considered the environment causing abiotic stresses. Under such situations, drought/flood tolerant genotypes can be identified through multi-location testing (MLT). Biotic stresses are a major limitation to plant productivity. Differences in disease and insect resistance can be associated with stable or unstable performance across environments. Multi-location testing helps in identifying the genotypes tolerant to particular disease/insect-pest provided the genotypes are grown in hot-spot locations under MLTs. As a matter of fact, the seed companies do select certain hot-spot locations just to screen the genotypes against major diseases under field conditions.

Breeding for Wider Adaptability: If a variety is in commercial cultivation over wide areas for more than 10-15 years, the variety is said to have wider adaptability and should be used as one of the parents in more number of crosses. In self-pollinated crops, a variety of proven adaptability is selected and used as a parent in the crossing programme and pedigree method of selection is commonly practiced for extracting recombinant pure-breeding lines, superior in yield and stability. The segregating generation population is raised in a number of contrasting environments alternately, for example, F_2 at location L1 and F_3 generation at contrasting location L2 and again F_4 generation at location L1 and so on. This method of breeding uses the principle of disruptive selection where the selection is practiced for the genotypes showing extreme phenotypes at the same time. International Maize and Wheat Improvement Centre (CIMMYT), Mexico called this method of breeding as "shuttle breeding". Here the individual/family having plasticity in different traits resulting into stability of yield over different environment is selected. Individual/family is selected on the basis of mean yield performance in such contrasting environment and advanced. The development of an individual is thus channeled into one optimal phenotype or another by a developmental switch mechanism (genetic/environmental). The general adaptability can thus also be attributed to variation in phenotypic plasticity.

In cross-pollinated crops, or self-pollinated crops showing good heterosis while developing single cross hybrids, inbreds/parental lines should be thoroughly evaluated and only those inbreds having higher adaptability should be used as parents. Here too, while developing lines/parents, principles of shuttle breeding may be taken into account to the extent possible.

Molecular approaches related to QTLs associated with adaptability can be adopted wherever, validated markers are available.

Realizing the importance of conduct of multi-location trials including a few hot-spot locations, private seed companies have created an independent unit within the R&D umbrella exclusively for this task and quite substantial resources are now allocated to this unit. ICAR itself runs a huge network of multi-location and multi-year testing of products developed by ICAR institutes and the state agricultural universities under crop specific All India Coordinated Research Projects.

Molecular Markers Related Programme

Now there is an increasing role of molecular markers in genotyping/DNA finger-printing of parental lines/hybrids, deployment of molecular assisted selection (MAS) especially in disease resistance breeding, molecular aided back-crossing (MAB) to introgress desirable genes from wild germplasm/ distantly related donors for disease resistance breeding and facilitate foreground and background selection in a typical back-crossing programme so as to recover the desired genotype by BC_3 instead of going all the way up to BC_6. The effective deployment of molecular markers adds speed and precision to the on-going regular breeding programme. The molecular breeders are not crop-wise instead they are across the crops and operate a well equipped lab to handle molecular markers related work. For small scale work, some companies prefer to out-source work related to molecular markers which sometimes becomes cost-effective.

R&D Operations

In bigger seed companies especially in the MNCs, the field operations and related activities like parental seed increase, maintenance of GMS and CMS lines, rapid generation advancement in greenhouse, double haploid production in the lab, handling of pre-breeding programmes and germplasm enhancement, procurement of inputs, labour management, devising ways to cut cost, data base management and all other operations important but not touched by the crop breeders are the responsibility of this division. This is headed by a professional plant breeder who can effectively communicate with the crop breeders and the testing team to add speed to the overall process of breeding to ensure that the products are delivered in the minimum possible time utilizing the resources in most effective manner.

Departments Not Under R&D But Fully Integrated with R&D

- Finance-Looks after financial matters.
- Human Resource (HR)-Looks after recruitment, training, job description, salary disbursement of employees and arranging their vehicles/laptops/ mobiles, etc.
- Commercial Seed Production at Farmers' field.
- Grow-out Test.
- Product Development-A coordinating type of department closely working with R&D, sales and marketing and having a keen eye on the products in the marketing and the choice of the consumers and

the farmers and giving feed-back to the R&D team to make strategic innovations in the products and also to have knowledge on sales price of the competitors' products and their positioning and acceptance and to have a good liaison with the farmers and sales team and the dealers and retailers.

- Seed Processing and Packaging-Deals with seed processing (conditioning) and packaging with attractive stickers and relevant details.
- Supply Chain Management-Deals with logistics, ware-housing and transport of the seed and other inputs as applicable from processing plant/ware-house to the dealer and retailers and to the farmers.
- Sales and Marketing-Deals with sales and marketing of the seeds and other inputs as per company guidelines, closely works with farmers and area dealer and retailers with sole objective to ensure that the seed and other saleable items of the company are sold as soon as possible with minimum of sales return.

4

Release and Notification of Field Crops Varieties in India

Varietal release and notification comes under the purview of the Seeds Act which aims to provide for regulating the quality of certain notified seeds for sale and for marketing and for the matters connected therewith. Released and notified varieties are offered for seed certification by the seed certification agencies which ensure that the varieties under certification meet certain prescribed field and seed standards failing which they stand to be rejected during the process of seed certification.

Purpose of Release

The purpose of Release of cultivars is to introduce the newly evolved varieties to the public and the farmers in particular for general cultivation in the region for which the variety is released and notified ensuring availability of seed of varieties according to their suitability and adaptation and avoiding mismatch. In other words, release of a cultivar is in the nature of a recommendation to the farmers for its adoption. Therefore, notification of a variety is linked to the release of a variety, though the release process itself does not have statutory cover.

Notification of Cultivars

After official release (at State as well and/or Central levels), the cultivars are notified under the Seeds Act so that the quality of seeds can be regulated. The main purpose of notification is to bring the seeds of a particular crop/ variety under the purview of Seed Law Enforcement, mainly to empower the Seed Inspectors to verify the quality of seeds by sampling and analysis. The notification regarding field crops varieties is made by the Central Government on the recommendation of the Central Sub-Committee on Crop Standards, Notification, and Release (CSCSNR) under the Chairmanship of ICAR-Deputy Director General (Crop Science).

The proposals for notification of a State released variety are forwarded in the prescribed format by the State government after its release in that State to the Central Seed Sub-Committee for consideration for notification. It is made clear that state can only release a variety but cannot notify the same as notification is a statutory process vested only in the central government which exercises this power through Ministry of Agriculture and Farmers' Welfare. Notification means publishing certain information in the Gazette of Government of India which is a public journal and an authorized legal document of government of India published weekly by the Department of Publication, Ministry of Housing and Urban Affairs. As a public journal, the Gazette prints official notices from the Government of India from time to time. In notification, an S. O. E. (Statutory Order, Extra-ordinary) is allotted to the group of varieties which are covered under that order.

Release of the Cultivars at Central and State Levels

The practice of official release of cultivars started in October, 1964 with the formation of the Central Variety Release Committee (CVRC) at the Central government level and State Variety Release Committee at State Government level. The Central Variety Release Committee functioned up to 1969 when its functions were taken over by the Central Seed Committee (CSC) established under the Seed Act, 1966. The Central Seeds Committee constituted a Central Sub-Committee on Crop Standards Notification and Release of Varieties for Agricultural Crops and a similar and separate committee for Horticultural Crops to discharge the functions of release/notification, provisional notification and de-notification of cultivars at Central level, while State Seed Sub-Committees (SSSC) were asked to discharge similar functions for release at State level. Initially, there was only one central variety release committee for all the crops covering field and horticultural crops. However, later on this committee was bifurcated into two and two independent committees were notified. Central Sub-Committee on Crop Standards, Notification and Release of Varieties for Agricultural Crops is responsible for release and notification of field crops. This committee is chaired by ICAR-Deputy Director General (Crop Science). The members are drawn from Directors/Crop Coordinators of ICAR, Directors Agriculture from States, Directors Seed Certification Agencies, Managing Directors of State Seed Corporations, Representatives from Private Seed Companies, etc. The crops under the purview of this committee are common cereals, pulses, oilseeds, millets, forage cash crops and fibre crops. Central Sub-Committee on Crop Standards, Notification and Release of Varieties for Horticultural Crops chaired by ICAR-Deputy Director General (Horticulture) takes care of horticultural crops.

The Deputy Commissioner (Quality Control) under Government of India Ministry of Agriculture and Farmers Welfare (Department of Agriculture, Cooperation and Farmers Welfare), F1211, Shastri Bhavan, New Delhi is the Member Secretary of the committee to receive the release and notification proposals from the sponsoring agencies, to prepare the agenda of the meeting, to convene the meeting and to prepare and circulate the minutes of the meetings.

It is necessary that all the proposals are complete in respect of enclosures viz., acknowledgement certificate regarding submission of seed sample with National Bureau of Plant Genetic Resources (NBPGR), New Delhi, detailed morphological characters of varieties, all India coordinated research project (AICRP) data on yield, maturity, disease and pests reaction, source of material and pedigree, package of practices of the cultivar, DNA fingerprinting profile, photographs on glossy sheet etc. Now it has been decided that the proposals which are not verified by the Director/Project Coordinator will not be considered by the committee. Hence submission of one copy of the proposal to the Director/Project Coordinator of the concerned crop is mandatory for considering the release and notification proposals. The better option would be to get the endorsement of project coordinator on the release proposal itself.

Currently 40 hard copies of the release/notification proposal along with necessary enclosures are submitted to the Dy. Commissioner Quality Control (QC), located at Shashtri Bhavan, New Delhi for further action at his secretariat. However, now efforts are on for development of portal for online submission of the varietal release and notification proposals to save time as well saving of stationary and postage. Release through this channel is open to all the organizations engaged in development of commercial cultivars including private sector seed companies. Private sector seed companies have to enter their cultivars/hybrids in the all India coordinated trials after paying certain prescribed testing fee for getting their products identified by the crop group meeting, released and notified by the designated agency.

Release vs. Notification

The Release is not a statutory function whose main purpose is to make known the details of the newly evolved cultivar to the public and also the areas for which it is adapted for cultivation based on its performance in all India coordinated trials/state regional research station trials/SAUs trials. The Notification is a statutory function performed under the Seeds Act-1966 so that the provisions of the Seeds Act could be stringently applied to regulate the quality of seeds during production, processing and sale and the offenders could be penalized by way of rejecting the seed not meeting the certain prescribed minimum field and seed standards where genetic purity of the seed and its germination are of prime concern.

Under the Seeds Act-1966, certified seeds can be produced only of notified varieties and thus, notification precedes the certification. Unless the variety is notified and the notification number along with document is shown to the seed certification inspector, he will refuse to proceed with the process of seed certification. Therefore, notification is compulsory for production of certified seeds. Seed Law Enforcement agency notified under the Seeds Act-1966 can draw and test samples of seeds of notified varieties. Therefore, for regulation of the quality of seeds, notification of the released cultivars either by the central variety release committee or state variety release committee is a precondition and notification for all kinds of releases related to field crops will be done only by Central Sub-Committee on Crop Standards, Notification and Release of Varieties for Agricultural Crops. The farmers benefit from the notification of varieties because they get seeds of assured quality of notified varieties. Seed quality is assured through the process of seed certification by state seed certification agencies meant for this purpose only. As specified under the Seeds Act, seeds of notified varieties can be sold after proper labeling and packing indicating the minimum specified seed standards. Once the variety is notified, in general, it becomes government of India asset. The morphological characters of notified varieties are documented by the Central Seed Committee so as to curtail the bio-piracy. Subsidies on seeds are considered based on the notification status. Notification facilitates the seed supply of designated varieties and hybrids of different crops by private sector seed companies to the state governments through the process of competitive bidding, etc. However, non-notified varieties are also sold in the market under TL (truthfully labeled) category and this helps the private seed companies whose most of the cultivars and the hybrids are not officially released and notified. However, these companies ensure seed quality and purity in their own interest and to protect their brand image and the seed standards are properly labeled on the seed packets/bags and they cannot escape the responsibility should the seed fail to meet the required quality and purity.

Procedure for Release of Cultivars

The varieties which are identified by the all India coordinated research project of the concerned crop or group of the crops are eligible for release and notification by Central Sub-committee on Crop Standards, Notification and Release of Varieties for Agricultural Crops. However, the cultivars which fail to be identified by the coordinated crop group meetings but perform well only in one State are eligible for release by the particular state variety release committee. These particular varieties which may have been evolved either by State Agricultural Universities/ICAR Institutes/Private Sector Seed Companies are to be considered by the State Seed Sub-committee of that particular State for release chaired by Director Agriculture of the State. The sponsoring authority intending to release the variety should furnish the relevant information in the prescribed pro-forma given as below and submit to the co-convener of the State Seed Sub-committee for consideration and release of cultivar. The varieties released by a particular state have to be notified by the Central Sub-Committee on Crop Standards, Notification and Release of Varieties for Agricultural Crops subsequently. The pro-forma for submission of proposals to central sub-committee or state seed sub-committee is given below. However, these need to be checked regularly and updated with the concerned crop coordinator/project director to use the pro-forma as applicable at a particular time.

Pro-forma for Submission of Proposal of Release of Crop Variety to Central Sub-Committee on Crop Standards, Notification and Release of Varieties (Central Varieties) as Applicable to All Crops

The pro-forma is as follows.

1	Name of the crop and species	:	
2	a) Name of the variety under which tested in AICRP trials	:	
	b) Proposed name of the Variety	:	
3	Sponsored by (institute)	:	
4	a) Institution or agency responsible for developing variety (with full address)	:	
	b) Name of the person who helped in the development of the variety	:	
	Developers	:	
	Collaborators	:	
5	a) Parentage (with details of its pedigree including source from which variety/inbred/ A, B and R lines of hybrid have been developed)	:	

	b) Source of material in case of introduction	:	
	c) DNA profile of variety/hybrid/ inbred/A, B, R line of hybrid vis-à-vis check variety/ line	:	
	d) Breeding method used	:	
	e) Breeding objective	:	
6	State the varieties which most closely resemble the proposed variety in general characters	:	
7	Recommended productions ecology (rain-fed/irrigated; high/ low fertility; season)	:	
8	Specific area of its adaptation (zones and states for which variety is proposed) and recommended productions ecology	:	
9	Description of hybrid/variety	:	
	a) All possible morphological characters	:	
	b) Distinguishing morphological characters	:	
	c) Maturity (range in number of days, from seedling/ transplanting to flowering, seed to seed)	:	
	d) Maturity group (early, medium and late wherever such classification exists)	:	
	e) Reaction to major diseases under field and controlled conditions (reaction to physiological strains/ races/ pathotypes/ bio-types to be indicated wherever possible)	:	
	f) Reaction to major pests (under field and controlled condition including store pests wherever applicable)	:	
	g) Agronomic features (e.g. resistance to lodging, shattering, fertilizer responsiveness, suitability to early or late sown conditions, seed rate etc.)	:	

	h) Quality of produce	:			
	Grain quality				
	Fodder quality				
	i) Reaction to stresses	:			
10	Description of the parents of the hybrid	:	A line/ Inbred 1	B line/Inbred 2	R line
	a) Plant height (cm)	:			
	b) Distinguishing morphological characters including grain characters	:			
	c) Days to flowering	:			
	d) Days to maturity (range in number of days – from seed to seed)	:			
	e) Is there any problem of synchronization? If yes, method to overcome it	:			
	f) Reaction to major diseases (under field and controlled conditions, reaction to physiological strains/ races/bio-types/ pathotypes to be indicated wherever possible)	:			
	g) Reaction to major pests (under field and controlled conditions including store pests)	:			
	h) Agronomic features (e.g. resistance to lodging, shattering, fertilizer responsiveness, suitability to early or late sown conditions, seed rate etc.)	:			
	i) Reaction to stresses	:			
11	a) Yield data in coordinated trials (breeding, agronomy, pathology, entomology, quality etc) regional/ inter regional district trials year wise (levels of fertilizer application, density of plant population and superiority over local control/standard variety to be indicated (to be attached)	:			

	b) Yield data from national, demonstration/large scale demonstrations (to be attached)	:			
12	a) Agency responsible for maintaining breeder seed	:			
	b) Quantity of breeder seed in stock (kg)	:			
	Variety				
	A line				
	B line				
	R line				
	Hybrid				
13	Specific recommendations, if any, for seed production (e.g. staggered sowing, plating ratio of parental lines of hybrids in foundation and certified seed production, probable area of seed production)	:			
14	Vivid presentation (field view, close-up of single plant and seed/ economic parts)	:			
15	a) Whether recommended by any workshop, seminar, conference, state seed committee etc.	:			
	b) If so, its recommendations with specific justifications for the release of proposed Variety				
16	c) Specific area of its adaptation	:			
17	Acknowledgement of submission of seed sample of variety/hybrid/ inbred/ A, B and R lines of hybrid from NBPGR and IC numbers	:			
18	Package of practices along with attainable yield levels	:			
19	Information on acceptability of the variety by farmers/ consumers/ industry	:			
20	Any other pertinent Information	:			

Signature of Breeder

Signature of Head of Institution

Checklist for Pro-forma for Submission of Proposal for Release of Crop - Variety to Central Sub-Committee on Crop Standards Notification and Release of Varieties

Details/Documents	Attached	
Parentage with details of its pedigree including source from which variety/Inbred/A, B and R lines of hybrid has been developed.	Yes	No
Source of material in case of introduction (IC/EC numbers provided by NBPGR).	Yes	No
Flow chart of details of development of variety/ parental lines of hybrids.	Yes	No
Molecular/ DNA profile of variety/hybrid/A, B, R line of hybrid vis-à-vis check variety/ line (details of unique amplicons with distinguishing markers along with photographs.	Yes	No
Detailed description of hybrid/variety.	Yes	No
Detailed description of the parental lines of hybrid.	Yes	No
Yield data and other data on diseases, insect-pest, quality etc. from coordinated trials.	Yes	No
Yield data from national, demonstration/large scale demonstrations.	Yes	No
Specific recommendations, if any, for seed production (e.g. staggered sowing, plating ratio of parental lines of hybrids in foundation and certified seed production, probable area of seed production etc.).	Yes	No
Vivid presentation (field view, close-up of single plant and seed) with the help of photographs of the variety).	Yes	No
Recommendation of workshop, conference.	Yes	No
Acknowledgement of submission of seed sample of variety/hybrid/ A, B and R lines of hybrid submitted to NBPGR.	Yes	No
Package of practices.	Yes	No
Pro-forma signed by all breeders and Head of Organization.	Yes	No
Any other pertinent information.	Yes	No

Standard Operating Procedure (SOP) for Release and Notification of Crop Varieties

1. Enter the variety/hybrid in the all India coordinated research project trial of the concerned crop by participating in the annual all India coordinated research project group meeting. While nominating the product, one has to submit station trial data for two years to demonstrate that the proposed variety or hybrid has some superiority over the existing check(s). Also ensure that the names of the proposed entries are included in the technical programme formulated during the workshop.

2. In general, there is a three tier testing in the coordinated trials of most of crops where the first year trial is designated as Initial Evaluation Trial (IET), the second year trial as Advance Varietal Trial I (AVT I) and the

third year trial as Advance Varietal Trial II (AVT II). Supply the seed as per details in the technical programme to the crop Coordinator well before the dead line.

3. Private sector seed companies must be prepared to deposit testing fee @ Rs.60,000/entry/year to the Project Coordinator as and when demanded to ensure that the testing data is made available to the sponsor. This fee keeps on changing and is usually increased after certain years. Non-compliance with depositing fee in time may lead to withdrawal of the entry from testing or if tested holding back the trial data.
4. Any variety/hybrid which has been successfully evaluated for three years in the coordinated trials, becomes eligible for identification by the annual coordinated crop workshop based on three years data and once the product has been identified, it becomes eligible for release and notification by Central Seed Sub-Committee on Crop Standards, Notification and Release for Agricultural Crops under the Chairmanship of ICAR-Deputy Director General (Crop Sciences).
5. Ensure to have a copy of the minutes/proceedings of the identification committee meeting showing that the particular variety/hybrid has been identified.
6. Submit required quantity of seed in the prescribed manner to the National Bureau of Plant Genetic Resources (NBPGR), New Delhi to obtain proper acknowledgement with National Identity number mentioning that the required seed has been deposited with the NBPGR for long term conservation. While submitting the seed to NBPGR, ensure the following compliances:
   - Deposit at least 5000 seeds in cross pollinated crops and at least 3000 seeds in self pollinated crops.
   - Seed should be freshly harvested and not more than 90 days old.
   - Seed should be clean, sound, healthy, and fully mature and harvested from healthy plants.
   - Potential viability of the seed should be more than 85 % in most crops except in cotton and vegetables where 70-80 % germination is acceptable.
   - Seed should not be treated with chemicals.
   - Seed should be packed in a good quality paper/muslin cloth/plastic packet with proper identity and preferably be repacked in a card board box to minimize damage and moisture absorption during transit.

7. After obtaining national identity number and acknowledgement from NBPGR, proceed to prepare the release proposal as per prescribed pro-forma for submission to Deputy Commissioner (QC), DAC, Shastri Bhavan, New Delhi (40 copies) along with following enclosures with each copy of proposal:
   - Relevant Minutes/relevant page of proceedings of the AICVIP-annual workshop showing that the variety/hybrid has been identified for release in the annual crop workshop. Preferably, this should be endorsed by Project Coordinator/Project Director.
   - Acknowledgement of receipt of seed by NBPGR with national identity number.
   - Glassy photograph of the variety/hybrid in A4 size.
   - Package of practices of the cultivar.
8. While preparing the release proposal, following points must be taken into consideration:
   - Source and genesis of parentage should be elaborated in detail.
   - The yield data should be presented year-wise where locations should be included under each year.
   - Each yield table (IET, AVT-I and AVT-II) should have marginal means, CD and CV, frequency of occurrence, and percentage increase or decrease over the checks.
   - There should be a summary yield table combining all the years and locations, again with CD, CV, mean, weighted mean, frequency of occurrence, percentage increase or decrease over the checks.
   - Include adequate data from coordinated trials on reaction to diseases and insect pests, and abiotic stresses.
   - Include data on grain quality.
9. Send 40 copies of the proposal complete in all respect with all relevant enclosures to Deputy Commissioner (Quality Control), Department of Agriculture and Cooperation (DAC) with one copy to Crop Project Coordinator-All India Coordinated Research Project.
10. Follow the progress regularly through appropriate correspondence with concerned authorities and visit the relevant website frequently (www.seednet.gov.in).

11. The sponsoring breeder may be called to make a power point presentation about the proposed cultivar in the central seed sub-committee meeting.
12. Once the cultivar is recommended for release and notification by the central seed sub-committee, one should look for minutes of the meeting on www.seednet.gov.in.
13. Take a print out of the minutes as a documentary proof of the recommendation for release and notification of the cultivar.
14. After release, there is Gazette Notification under certain Statutory Order (SO) number and date and that notification also appears on the seed portal website www.seednet.gov.in and this happens a few months later after the release and usually before the next meeting on release and notification.
15. Take a print-out of this notification and file for reference and official use.
16. The release proposal of varieties which are not identified for release by the concerned crop workshop but have done exceptionally well in a particular state can be routed through Director-Agriculture of that particular state. In this situation, the cultivar is further tested in the state at regional research stations and the complete proposal having station trial data, coordinated trial data and state trial data is put-up to Director Agriculture for consideration and release in that particular state. Once the proposal is approved for release by the state variety release committee, then the entire release proposal (40 copies) along with recommendation of the state variety release committee meeting is again sent to Deputy Commissioner (QC), DAC, New Delhi with copies to concerned Crop Project-Coordinator for notification. During this submission, the requirements as mentioned under SN 7 and 8 must be met and there should be strong follow-up action as mentioned in case of release and notification of cultivars by central seed sub-committee (Ram, 2020).

A few Possible Causes of Rejection of Notification Proposals of the State Variety Release Committees

- No DNA fingerprinting profile.
- No glossy colour photo in A4 size.
- High disease severity.
- No details on package of practices.

- Not sufficient data on nutritional contents if the varieties are claimed to be bio-fortified.
- Lack of AICRIP data particularly on reaction to diseases and insect-pests.
- Varietal ownership dispute.
- Non-submission of copy of minutes of state seed sub-committee.
- Long gap years (about more than 4 years between the date of release by the state seed sub-committee and the submission of notification proposal to the central seed sub-committee.

5

Seed Act-1966

The Seed Act 1966 was passed by the Indian Parliament on 29th December 1966 and it came into force from 2nd October, 1969. In addition, Seed Regulation Order was also passed in December 1983. The sole objective of Seed Act and Order has been to ensure that the Indian farmers get good quality seeds of important crops. Important provisions and guidelines for the farmers are given below:

1. The Act is applicable for the seeds notified by the Central Government from time to time.
2. Seed quality is regulated at two stages viz. seed production and seed certification. For this purpose, seed testing laboratories and seed certification agencies are to be established in each state. Now all the states have implemented this act.

Sale of Seeds

1. Any institution, agency, company or individual wishing to do business in seeds must obtain necessary permission for buying and selling seeds.
2. License holder must keep all seeds notified by the Central Govt. for sale.
3. The seeds to be sold to farmers must satisfy the minimum quality in terms of standards of physical purity and germination percentage fixed under the Seed Act.
4. The bags or packets containing seeds for sale must have specifications fixed for that particular seed. These labels are of different colours for different seeds for the convenience of buying farmers.

   i. Breeder's seed - Yellow colour

   ii. Foundation seed - White colour

   iii. Certified seed - Blue colour

1. The labels must contain following information on them.
    i. Seed - Name and variety
    ii. Lot number
    iii. Physical purity (in percent) and germination (in percent)
    iv. Date of testing
    v. Weight of the seed in the bag/packet
    vi. Seed class/category.
    vii. Expiry date/viability period
    viii. Name and address of seed producer.
    ix. Signature of the Officer of seed certification agency.

Care by Farmers

Farmers should take following care while purchasing the seeds.

1. They should buy required seed from the authorized license holders only.
2. They should obtain printed receipt from the dealer/shopkeeper for the seed purchased and payment made. The receipt should contain important specifications of the seed written on the label viz. name of the buyer, name of the crop and variety, plot numbers, producer's name, signature of shopkeeper, dealer, etc.
3. While buying the seed they should see carefully all the entries on the label particularly germination, percentage, date of seed test, etc.
4. They should ensure that the label is duly signed by the officer of Seed Certification Agency.
5. The bag should be opened from the button side keeping the stitches of the upper side and the label intact. Empty bags with tag and the printed receipt should be preserved for seasonal period.
6. If there are two or more bags of seed, the seed from them should be sown separately without mixing.
7. The seed should be sown when there is optimum moisture and the date of sowing should be recorded.
8. If there is any complaint about the quality of seed, the matter should be reported to Seed Inspector or to Agricultural Development Officer of

Zilla Parishad of the concerned district. If there is financial loss to the farmer, he can approach the District Consumer Forum for compensation.

Seed Quality Standards

The physical purity and germination percentage of some seeds as a measure of quality standard as prescribed by Seed Act-1966 are given in Table 5.1 (www.seednet.gov.in).

Table 5.1: Physical purity and germination standards in selected field crops as per seed act-1966

S.N.	Crop	Physical purity (%)	Germination (%)
1	Jowar and Bajra	98	75
2	Wheat	98	85
3	Rice, Maize	98	80
4	Groundnut	96	70
5	Soyabean	97	70
6	Sunflower	97	60
7	Sesamum	97	80
8	Gram	98	85
9	Pigeon Pea, Green Gram, Black Gram	98	75
10	Cotton	98	65

Seed Act-1966

The essential features of Indian Seed Act-1966 (Act No. 54 of 1966) are as follows.

Short Title, Extent and Commencement

1.(1) This Act may be called the Seeds Act, 1966.

(2) It extends to the whole of India.

(3) It shall come into force on such date as the Central Government may, by notification in the Official Gazette, appoint, and different dates may be appointed for different provisions of this Act, and for different States or for different areas thereof.

Definitions

2. In this Act, unless the context otherwise requires,

(1) "Agriculture" includes horticulture;

(2) "Central Seed Laboratory" means the Central Seed Laboratory established or declared as such under sub-section (1) of section 4;

(3) "Certification agency" means the certification agency established

(4) "Committee" means the Central Seed Committee constituted under sub-section (1) of Section 3;

(5) "Container" means a box, bottle, casket, tin, barrel, case, receptacle, sack, bag, wrapper or other thing in which any article or thing is placed or packed;

(6) "Export" means taking out of India to a place outside India;

(7) "Import" means bringing into India from a place outside India;

(8) "Kind" means one or more related species or sub-species of crop plants each individually or collectively known by one common name such as cabbage, maize, paddy and wheat;

(9) "notified kind or variety", in relation to any seed, means any kind or variety thereof notified under Section 5;

(10) "prescribed" means prescribed by rules made under this act;

(11) "seed" means any of the following classes of seeds used for sowing or planting-

  (i) seeds of food crops including edible oil seeds and seeds of fruits and vegetables;

  (ii) cotton seeds;

  (iii) seeds of cattle fodder; and includes seedlings, and tubers, bulbs, rhizomes, roots, cuttings, all types of grafts and other vegetatively propagated material, of food crops or cattle fodder;

(12) "Seed Analyst" means a Seed Analyst appointed under section 12;

(13) "Seed Inspector" means a Seed Inspector appointed under section 13;

(14) "State Government", in relation to a Union territory, means the administrator thereof;

(15) "State Seed Laboratory", in relation to any State, means the State Seed Laboratory established or declared as such under sub-section (2) of section 4 for that State; and

(16) "variety" means a sub-division of a kind identifiable by growth, yield, plant, fruit, seed, or other characteristic.

Central Seed Committee

3.(1) The Central Government shall, as soon as may be after the commencement of this Act, constitute a Committee called the Central Seed Committee to advise the Central Government and the State Governments on matters arising out of the administration of this Act and to carry out the other functions assigned to it by or under this Act.

(2) The Committee shall consist of the following members, namely:

(i) a Chairman to be nominated by the Central Government;

(ii) eight persons to be nominated by the Central Government to represent such interests that Government thinks fit, of whom not less than two persons shall be representatives of growers of seed;

(iii) one person to be nominated by the Government of each of the States.

(3) The members of the Committee shall, unless their seats become vacant earlier by resignation, death or otherwise, be entitled to hold office for two years and shall be eligible for re-nomination.

(4) The Committee may, subject to the previous approval of the Central Government, make bye-laws fixing the quorum and regulating its own procedure and the conduct of all business to be transacted by it.

(5) The Committee may appoint one or more sub-committees, consisting wholly of members of the Committee or wholly of other persons or partly of members of the Committee and partly of other persons, as it thinks fit, for the purpose of discharging such of its functions as may be delegated to such sub-committee or sub-committees by the Committee.

(6) The functions of the Committee or any sub-committee thereof may be exercised notwithstanding any vacancy therein.

(7) The Central Government shall appoint a person to be the secretary of the Committee and shall provide the Committee with such clerical and other staff as the Central Government considers necessary.

Central Seed Laboratory and State Seed Laboratory

4.(1) The Central Government may, by notification in the Official Gazette, establish a Central Seed Laboratory or declare any seed laboratory as the Central Seed Laboratory to carry out the functions entrusted to the Central Seed Laboratory by or under this Act.

(2) The State Government may, by notification in the Official Gazette, establish one or more State Seed Laboratories or declare any seed laboratory as a State Seed Laboratory where analysis of seeds of any notified kind or variety shall be carried out by Seed Analysts under this Act in the prescribed manner.

Power to Notify Kinds or Varieties of Seeds

5. If the Central Government, after consultation with the committee, is of opinion that it is necessary or expedient to regulate the quality of seed of any kind or variety to be sold for purposes of agriculture, it may, by notification in the Official Gazette, declare such kind or variety to be a notified kind or variety for the purposes of the Act and different kinds or varieties may be notified for different States or for different areas thereof.

Power to Specify Minimum Limits of Germination and Purity, etc.

6. The Central Government may, after consultation of the Committee and by notification in the Official Gazette, specify –

   (a) the minimum limits of germination and purity with respect to any seed of any notified kind or variety;

   (b) the mark or label to indicate that such seed conforms to the minimum limits of germination and purity specified under clause (a) and the particulars which marks or label may contain.

Regulation of Sale of Seeds of Notified Kinds or Varieties

7. No person shall, himself or by any other person on his behalf, carry on the business of selling, keeping for sale, offering to sell, bartering or otherwise supplying any seed of any notified kind or variety, unless-

   (a) such seed is identifiable as to its kind or variety;

   (b) such seed conforms to the minimum limits of germination and purity specified under clause (a) of section 6;

   (c) the container of such seed bears in the prescribed manner, the mark or label containing the correct particulars thereof, specified under clause (b) of section 6; and

   (d) he complies with such other requirements as may be prescribed.

Seed Certification Agency

8. The State Government or the Central Government in consultation with the State Government may, by notification in the Official Gazette, establish a certification agency for the State to carry out the functions entrusted to the certification agency by or under this Act.

Grant of Certificate by Certification Agency

9.(1) Any person selling, keeping for sale, offering to sell, bartering or otherwise supplying any seed of any notified kind or variety may, if he desires to have such seed certified by the certification agency, apply to the certification agency for the grant of a certificate for the purpose.

(2) Every application under sub-section (1) shall be made in such form, shall contain such particulars and shall be accompanied by such fees as may be prescribed.

(3) On receipt of any such application for the grant of a certificate, the certification agency may, after such enquiry as it thinks fit and after satisfying itself that the seed to which the application relates conforms to the minimum limits of germination and purity specified for that seed under clause (a) of section 6, grant a certificate in such form and on such conditions as may be prescribed.

Revocation of Certificate

10. If the certification agency is satisfied, either on a reference made to it in this behalf or otherwise, that-

(a) the certificate granted by it under section 9 has been obtained by misrepresentation as to an essential fact; or

(b) the holder of the certificate has, without reasonable cause, failed to comply with the conditions subject to which the certificate has been granted or has contravened any of the provisions of this Act or the rules made thereunder; then, without prejudice to any other penalty to which the holder of the certificate may be liable under this Act, the certification agency may, after giving the holder of the certificate an opportunity of showing cause, revoke the certificate.

Appeal

11.(1) Any person aggrieved by a decision of a certification agency under section 9 or section 10, may, within thirty days from the date on which the decision is communicated to him and on payment of such fees as may be prescribed, prefer an appeal to such authority as may be specified by the State Government in this behalf:

Provided that the appellate authority may entertain an appeal after the expiry of the said period of thirty days if it is satisfied that the appellate was prevented by sufficient cause from filing the appeal in time.

(2) On receipt of an appeal under sub-section (1), the appellate authority shall, after giving the appellant an opportunity of being heard, dispose of the appeal as expeditiously as possible.

(3) Every order of the appellate authority under this section shall be final.

Seed Analysts

12. The State Government may, by notification in the Official Gazette, appoint such persons as it thinks fit, having the prescribed qualifications, to be Seed Analysts and define the areas within which they shall exercise jurisdiction.

Seed Inspectors

13.(1)The State Government may, by notification in the Official Gazette, appoint such persons as it thinks fit, having the prescribed qualifications, to be Seed Inspectors and define the areas within which they shall exercise jurisdiction.

(2) Every Seed Inspector shall be deemed to be a public servant within the meaning of section 21 of the Indian Penal Code (45 of 1860) and shall be officially subordinate to such authority as the State Government may specify in this behalf.

Powers of Seed Inspector

14.(1)The Seed Inspector may-

(a) take samples of any seed of any notified kind or variety from-

(i) any person selling such seed; or

(ii) any person who is in the course of conveying, delivering or preparing to deliver such seed to a purchaser or a consignee; or

(iii) a purchaser or a consignee after delivery of such seed to him;

(b) send such sample for analysis to the Seed Analyst for the area within which such sample has been taken;

(c) enter and search at all reasonable times, with such assistance, if any, as he considers necessary, any place in which he has reason to believe that an offence under this Act has been or is being committed and order in writing the person in possession of any seed in respect of which the offence has been or is being committed, not to dispose of any stock of such seed for a specific period not exceeding thirty days or, unless the alleged offence is such that the defect may be removed by the possessor of the seed, seize the stock of such seed;

(d) examine any record, register, document or any other material object found in any place mentioned in clause (c) and seize the same if he has reason to believe that it may furnish evidence of the commission of an offence punishable under this Act; and

(e) exercise such other powers as may be necessary for carrying out the purposes of this Act or any rule made thereunder.

(2) Where any sample of any seed of any notified kind or variety is taken under clause (a) of sub-section (1), its cost, calculated at the rate at which such seed is usually sold to the public, shall be paid on demand to the person from whom it is taken.

(3) The power conferred by this section includes power to break-open any container in which any seed of any notified kind or variety may be contained or to break-open the door of any premises where any such seed may be kept for sale:

Provided that the power to break-open the door shall be exercised only after the owner or any other person in occupation of the premises, if he is present therein, refuses to open the door on being called upon to do so.

(4) Where the Seed Inspector takes any action under clause (a) of sub-section (1), he shall, as far as possible, call not less than two persons to be present at the time when such action is taken and take their signatures on a memorandum to be prepared in the prescribed form and manner.

(5) The provisions of the Code of Criminal Procedure, 1898 (5 of 1898), shall, so far as may be, apply to any search or seizure under this section as they apply to any search or seizure made under the authority of a warrant issued under section 98 of the said Code.

Procedure to be Followed by Seed Inspectors

15.(1)Whenever a Seed Inspector intends to take sample of any seed of any notified kind or variety for analysis, he shall-

(a) give notice in writing, then and there, of such intention to the person from whom he intends to take sample;

(b) except in special cases provided by rules made under this Act, take three representative samples in the prescribed manner and mark and seal or fasten up each sample in such manner as its nature permits.

(1) When samples of any seed of any notified kind or variety are taken under sub-section (1), the Seed Inspector shall-

(a) deliver one sample to the person from whom it has been taken;

(b) send in the prescribed manner another sample for analysis to the Seed Analyst for the area within which such sample has been taken; and

(c) retain the remaining sample in the prescribed manner for production in case any legal proceedings are taken or for analysis by the Central Seed Laboratory under sub-section (2) of section 16, as the case may be.

(2) If the person from whom the samples have been taken refuses to accept one of the samples, the Seed Inspector shall send intimation to the Seed Analyst of such refusal and thereupon the Seed Analyst receiving the sample for analysis shall divide it into two parts and shall seal or fasten up one of those parts and shall cause it, either upon receipt of the sample or when he delivers his report, to be delivered to the Seed Inspector who shall retain it for production in case legal proceedings are taken.

(3) Where a Seed Inspector takes any action under clause (c) of sub-section (1) of section 14:

(a) he shall use all dispatch in ascertaining whether or not the seed contravenes any of the provisions of section 7 and if it is ascertained that the seed does not so contravene, forthwith revoke the order passed under the said clause or, as the case may be, take such action as may be necessary for the return of the stock of the seed seized;

(b) if he seizes the stock of the seed, he shall, as soon as may be, inform a magistrate and take his orders as to the custody thereof;

(c) without prejudice to the institution of any prosecution, if the alleged offence is such that the defect may be removed by the possessor of the seed, he shall, on being satisfied that the defect has been so removed, forthwith revoke the order passed under the said clause.

(4) Whereas Seed Inspector seizes any record, register, document or any other material object under clause (d) of sub-section (1) of section 14, he shall, as soon as may be, inform a magistrate and take his orders as to the custody thereof.

Report of Seed Analyst

16.(1)The Seed Analyst shall, as soon as may be after the receipt of the sample under sub-section (2) of section 15, analyze the sample at the State Seed Laboratory and deliver, in such form as may be prescribed, one copy of the report of the result of the analysis to the Seed Inspector and another copy thereof to the person from whom the sample has been taken.

(2) After the institution of a prosecution under this Act, the accused vendor or the complainant may, on payment of the prescribed fee, make an application to the court for sending any of the samples mentioned in clause

(a) or clause (c) of sub-section (2) of section 15 to the Central Seed Laboratory for its report and on receipt of the application, the court shall first ascertain that the mark and the seal or fastening as provided in clause

(b) of sub-section (1) of section 15 are intact and may then dispatch the sample under its own seal to the Central Seed Laboratory which shall thereupon send its report to the court in the prescribed form within one month from the date of receipt of the sample, specifying the result of the analysis.

(3) The report sent by the Central Seed Laboratory under sub-section (2) shall supersede the report given by the Seed Analyst under sub-section (1).

(4) Where the report sent by the Central Seed Laboratory under sub-section (2) is produced in any proceedings under Section 19, it shall not be necessary in such proceedings to produce any sample or part thereof taken for analysis.

Restriction on Export and Import of Seeds of Notified Kinds or Varieties

17. No person shall, for the purpose of sowing or planting by any person (including himself), export or import or cause to be exported or imported any seed of any notified kind or variety, unless-

    (a) it conforms to the minimum limits of germination and purity specified for that seed under clause (a) of section 6; and

    (b) its container bears, in the prescribed manner, the mark or label with the correct particulars thereof specified for that seed under clause (b) of section 6.

Recognition of Seed Certification Agencies of Foreign Countries

18. The Central Govt. may, on the recommendation of the Committee and by notification in the Official Gazette, recognize any seed certification agency established in any foreign country, for the purposes of this Act.

Penalty

19. If any person-

(a) contravenes any provision of this Act or any rule made thereunder; or

(b) prevents a Seed Inspector from taking sample under this Act; or

(c) prevents a Seed Inspector from exercising any other power conferred on him by or under this Act; he shall, on conviction, be punishable-

    (i) for the first offence with fine which may extend to five hundred rupees, and

    (ii) in the event of such person having been previously convicted of an offence under this section, with imprisonment for a term which may extend to six months, or with fine which may extend to one thousand rupees, or with both.

Forfeiture of Property

20. When any person has been convicted under this Act for the contravention of any of the provisions of this Act or the rules made thereunder, the seed in respect of which the contravention has been committed may be forfeited to the Government.

Offences by Companies

21.(1) Where an offence under this Act has been committed by a company, every person who at the time the offence was committed was in charge of, and was responsible to the company for the conduct of the business of the company, as well as the company, shall be deemed to be guilty of the offence and shall be liable to be proceeded against and punished accordingly:

Provided that nothing contained in this sub-section shall render any such person liable to any punishment under this Act if he proves that the offence was committed without his knowledge and that he exercised all due diligence to prevent the commission of such offence.

(2) Notwithstanding anything contained in sub-section (1), where an offence under this Act has been committed by a company and it is proved that the offence has been committed with the consent or connivance of, or is attributable to any neglect on the part of, any director, manager, secretary or other officer of the company, such director, manager, secretary or other officer shall also be deemed to be guilty of that offence and shall be liable to be proceeded against and punished accordingly.

Protection of Action Taken in Good Faith

22. No suit, prosecution or other legal proceeding shall lie against the Government or any officer of the Government for anything which is in good faith done or intended to be done under this Act.

Power to Give Directions

23. The Central Government may give such directions to any State Government as may appear to the Central Government to be necessary for carrying into execution in the State any of the provisions of this Act or of any rule made thereunder.

Exemption

24. Nothing in this Act shall apply to any seed of any notified kind or variety grown by a person and sold or delivered by him on his own premises direct to another person for being used by that person for the purpose of sowing or planting.

Power to Make Rules

25.(1)The Central Government may, by notification in the Official Gazette, make rules to carry out the purpose of this Act.

(2) In particular and without prejudice to the generality of the fore-going power, such rules may provide, for-

(a) the functions of the Committee and the travelling and daily allowances payable to members of the Committee and members of any sub-committee appointed under sub-section (5) of section 3;

(b) the functions of the Central Seed Laboratory;

(c) the functions of a certification agency;

(d) the manner of marking or labeling the container of seed of any notified kind or variety under clause (c) of Section 7 and under clause (b) of section 17;

(e) the requirements which may be complied with by a person carrying on the business referred to in section 7;

(f) the form of application for the grant of a certificate under section 9, the particulars it may contain, the fees which should accompany it, the form of the certificate and the conditions subject to which the certificate may be granted;

(g) the form and manner in which and the fee on payment of which an appeal may be preferred under section 11 and the procedure to be followed by the appellate authority in disposing of the appeal;

(h) the qualifications and duties of Seed Analysts and Seed Inspectors;

(i) the manner in which samples may be taken by the Seed Inspector, the procedure for sending such samples to the Seed Analyst or the Central Seed Laboratory and the manner of analyzing such samples;

(j) the form of report of the result of the analysis under sub-section (1) or sub-section (2) of section 16 and the fees payable in respect of such report under the said sub-section (2);

(k) the records to be maintained by a person carrying on the business referred to in section 7 and the particulars which such records shall contain; and

(l) any other matter which is to be or may be prescribed.

(3) Every rule made under this Act shall be laid as soon as may be after it is made, before each House of Parliament while it is in session for a total period of thirty days which may be comprised in one session or in two successive sessions, and if, before the expiry of the session in which it is so laid or the session immediately following, both Houses agree in making any modification in the rule or both Houses agree that the rule should not be made, that rule shall, thereafter have effect only in such modified form or be of no effect, as the case may be; so however, that any such modification or annulment shall be without prejudice to the validity of anything previously done under that rule (www.seednet.gov.in).

6

Protection of Plant Varieties and Farmers' Rights Act-2001

Protection of Plant Varieties and Farmers' Rights Act-2001 (PPVFRA-2001) was enacted by government of India in 2001 to create a *sui generis* system of providing a legal framework as an IPR mechanism to protect the varieties of crop plants in India along with grant of breeders' and farmers' rights in the interest of all the stakeholders and also to comply with WTO guidelines to which India is a legal signatory. The salient features of this act are as follows.

The Protection of Plant Varieties and Farmers' Rights Act-2001, Act No. 53 of 2001, (30th October, 2001)

1. Short Title, Extent and Commencement

(1) This Act may be called the Protection of Plant Varieties and Farmers' Rights Act, 2001.

(2) It extends to the whole of India.

(3) It shall come into force on such date as the Central Government may, by notification in the Official Gazette, appoint; and different dates may be appointed for different provisions of this Act and any reference in any such provision to the commencement of this Act shall be construed as a reference to the coming into force of that provision.

2. Definitions: In this Act, unless the context otherwise requires,

(a) "Authority" means the Protection of Plant Varieties and Farmers' Rights Authority established under sub-section (1) of section 3;

(b) "benefit sharing", in relation to a variety, means such proportion of the benefit accruing to a breeder of such variety or such proportion of the benefit accruing to the breeder from an agent or a licensee of such variety, as the case may be, for which a claimant shall be entitled as determined by the Authority under section 26.

(c) "breeder" means a person or group of persons or a farmer or group of farmers or any institution which has bred, evolved or developed any variety;

(d) "convention country" means a country which has acceded to an international convention for the protection of plant varieties to which India has also acceded, or a country which has a law on protection of plant varieties on the basis of which India has entered into an agreement for granting plant breeders' right to the citizens of both the countries;

(e) "denomination", in relation to a variety or its propagating material or essentially derived variety or its propagating material, means the denomination of such variety or its propagating material or essentially derived variety or its propagating material, as the case may be, expressed by means of letters or a combination of letters and figures written in any language;

(f) "essential characteristics" means such heritable traits of a plant variety which are determined by the expression of one or more genes of other heritable determinants that contribute to the principle features, performance or value of the plant variety;

(g) "essentially derived variety", in respect of a variety (the initial variety), shall be said to be essentially derived from such initial variety when it—

  (i) is predominantly derived from such initial variety, or from a variety that itself is predominantly derived from such initial variety, while retaining the expression of the essential characteristics that results from the genotype or combination of genotype of such initial variety;

  (ii) is clearly distinguishable from such initial variety; and

  (iii) conforms (except for the differences which result from the act of derivation) to such initial variety in the expression of the essential characteristics that result from the genotype or combination of genotype of such initial variety;

(h) "extant variety" means a variety available in India which is-

  (i) notified under section 5 of the Seeds Act, 1966 or

  (ii) a variety about which there is common knowledge; or (iii) any other variety which is in public domain;

(i) "farmer" means any person who-(*i*) cultivates crops by cultivating the land himself; or (*ii*) cultivates crops by directly supervising the cultivation of land through any other person; or (*iii*) conserves and preserves, severally or jointly, with any person any wild species or traditional varieties or adds value to such wild species or traditional varieties through selection and identification of their useful properties;

(j) "farmers' variety" means a variety which-(*i*) has been traditionally cultivated and evolved by the farmers in their fields; or (*ii*) is a wild relative or land race of a variety about which the farmers possess the common knowledge;

(k) "Gene Fund" means the National Gene Fund constituted under sub-section (*1*) of section 45;

(l) "Propagating material" means any plant or its component or part thereof including an intended seed or seed which is capable of, or suitable for, regeneration into a plant;

(m) "Register" means a national Register of Plant Varieties referred to in section 13;

(n) "Registrar" means a Registrar of Plant Varieties appointed under sub-section (*4*) of section 12 and includes the Registrar-General;

(o) "Registrar-General" means the Registrar-General of Plant Varieties appointed under sub-section (*3*) of section12;

(p) "Registry" means the Plant Variety Registry referred to in sub-section (*1*) of section 2;

(q) "seed" means a type of living embryo or propagule capable of regeneration and giving rise to a plant which is true to such type;

(r) "variety", means a plant grouping except micro-organism within a single botanical taxon of the lowest known rank, which can be- (*i*) defined by the expression of the characteristics resulting from a given genotype of that plant grouping; (*ii*) distinguished from any other plant grouping by expression of at least one of the said characteristics; and (*iii*) considered as a unit with regard to its suitability for being propagated, which remains unchanged after such propagation, and includes propagating material of such variety, extant variety, transgenic variety, farmers' variety and essentially derived variety.

3. Establishment of Authority

(1) The Central Government shall, by notification in the Official Gazette, establish an authority to be known as the Protection of Plant Varieties and Farmers' Rights Authority for the purposes of this Act.

(2) The Authority shall be a body corporate by the name aforesaid, having perpetual succession and a common seal with power to acquire, hold and dispose of properties, both movable and immovable, and to contract, and shall by the said name sue and be sued.

(3) The head office of the Authority shall be at such place as the Central Government may, by notification in the Official Gazette, specify and the Authority may, with the previous approval of the Central Government, establish branch offices at other places in India.

(4) The Authority shall consist of a Chairperson and fifteen members.

(5) (a) The Chairperson to be appointed by the Central Government shall be a person of outstanding caliber and eminence, with long practical experience to the satisfaction of that Government especially in the field of plant varietal research or agricultural development.

(b) The members of the Authority, to be appointed by the Central Government, shall be as (*i*) the Agriculture Commissioner, Government of India, Department of Agriculture and Cooperation, New Delhi, *ex officio*; (*ii*) the Deputy Director General in-charge of Crop Sciences, Indian Council of Agricultural Research, New Delhi, *ex officio*; (*iii)* the Joint Secretary in-charge of Seeds, Government of India, Department of Agriculture and Cooperation, New Delhi, *ex officio*;(*iv*) the Horticulture Commissioner, Government of India, Department of Agriculture and Cooperation, New Delhi, *ex officio;* (*v*) the Director, National Bureau of Plant Genetic Resources, New Delhi, *ex officio*; (*vi*) one member not below the rank of Joint Secretary to the Government of India, to represent the Department of Bio-technology, Government of India, *ex officio*; (*vii*) one member not below the rank of Joint Secretary to the Government of India to represent the Ministry of Environment and Forests, Government of India, *ex officio*; (*viii*) one member not below the rank of Joint Secretary to the Government of India to represent the Ministry of Law, Justice and Company Affairs, Government of India, *ex officio*; (*ix*) one representative from a National or State level farmers' organization to be nominated by the Central Government; (*x*) one representative from a tribal organization to be nominated by the Central Government; (*xi*) one representative from the seed industry to be nominated by the Central Government; (*xii*) one representative from an agricultural University to be nominated by the Central Government; (*xiii*) one representative from a National or State level women's organization associated with agricultural activities to be nominated by the Central Government; and (*xiv*) two representatives of State Governments on rotation basis to be nominated by the Central Government.

(c) The Registrar-General shall be the *ex officio* member-secretary of the Authority.

(6) The term of office of the Chairperson and the manner of filling the post shall be such as may be prescribed.

(7) The Chairperson shall appoint a Standing Committee consisting of five members, one of whom shall be a member who is a representative from a farmers' organization, to advise the Authority on all issues including farmers' rights.

(8) The Chairperson shall be entitled to such salary and allowances and shall be subject to such conditions of service in respect of leave, pension, provident fund and other matters as may be prescribed.

4.Meetings of Authority

(1) The Authority shall meet at such time and place and shall observe such rules of procedure in regard to the transaction of business at its meetings [including the quorum at its meetings and the transaction or business of its Standing Committee appointed under sub-section (*7*) of section 3] as may be prescribed. (2) The Chairperson of the Authority shall preside at the meetings of the Authority.

5. Committees of Authority: (1) The Authority may appoint such committees as may be necessary for the efficient discharge of its duties and performance of its functions under this Act.

6. Officers and Other Employees of Authority: Subject to such control and restriction as may be prescribed, the Authority may appoint such officers and other employees as may be necessary for the efficient performance of its functions.

7. Chairperson to be Chief Executive: The Chairperson shall be the Chief Executive of the Authority and shall exercise such powers and perform such duties as may be prescribed.

8. General Functions of Authority: (1) It shall be the duty of the Authority to promote, by such measures as it thinks fit, the encouragement for the development of new varieties of plants and to protect the rights of the farmers and breeders. (2) In particular, and without prejudice to the generality of the foregoing provisions, the measures referred to in sub-section (1) may provide for (a) the registration of extant varieties subject to such terms and conditions and in the manner as may be prescribed; (b) developing characterisation and documentation of varieties registered under this Act; (c) documentation, indexing and cataloguing of farmers' varieties; (d) compulsory cataloguing facilities for all varieties of plants; (e) ensuring that seeds of the varieties registered under this Act are available to the farmers and providing for compulsory licensing of such varieties if the breeder of such varieties or any other person entitled to produce such variety under this Act does not arrange for production and sale of the seed in the manner as may be prescribed; (*f*) collecting statistics with regard to plant varieties, including the contribution of any person at any time in the evolution or development of any plant variety, in India or in any other country, for compilation and publication.

9. Authentication of Orders, Etc., of Authority: All orders and decisions of the Authority shall be authenticated by the signature of the Chairperson or any other member authorized by the Authority in this behalf.

10. Delegation: The Authority may, by general or special order in writing, delegate to the Chairperson, any member or officer of the Authority subject to such conditions or limitations, if any, as may be specified in the order, such of its powers and functions (except the power to make regulations under section 95) under this Act as it may deem necessary.

11. Power of Authority: In all proceedings under this Act before the Authority or the Registrar, (a) the Authority or the Registrar, as the case may be, shall have all the powers of a civil court for the purposes of receiving evidence, administering oaths, enforcing the attendance of witnesses, compelling the discovery and production of documents and issuing commissions for the examination of witnesses; (b) the Authority or the Registrar may, subject to any rules made in this behalf under this Act, make such orders as to cost as it considers reasonable and any such order shall be executable as a decree of a civil court.

12. Registry and Offices Thereof: (1) The Central Government shall establish, for the purpose of this Act, a Registry which shall be known as the Plant Varieties Registry. (2) The head office of the Plant Varieties Registry shall be located in the head office of the Authority. (3) The Authority shall appoint a Registrar-General of Plant Varieties who shall be entitled to such salary and allowances and shall be subject to such conditions of service in respect of leave, pension, provident fund and such other matters as may be prescribed. (4) The Authority may appoint such number of Registrars as it thinks necessary for registration of plant varieties under the superintendence and direction of the Registrar-General under this Act and may make regulations with respect to their duties and jurisdiction.

13. National Register of Plant Varieties: For the purposes of this Act, a Register called the National Register of Plant Varieties shall be kept at the head office of the Registry, wherein shall be entered the names of all the registered plant varieties with the names and addresses of their respective breeders, the right of such breeders in respect of the registered varieties, the particulars of the denomination of each Registered variety, its seed or other propagating material along with specification of salient features thereof and such other matters as may be prescribed.

14. Application for Registration: Any person specified in section 16 may make an application to the Registrar for registration of any variety: (*a*) of such genera and species as specified under sub-section (*2*) of section 29; or (*b*) which is an extant variety; or (*c*) which is a farmer's variety.

15. Registrable Varieties: (1) A new variety shall be registered under this Act if it conforms to the criteria of novelty, distinctiveness, uniformity and stability. (2) Notwithstanding anything contained in sub-section (1), an extant variety shall be registered under this Act within a specified period if it conforms to such criteria of distinctiveness, uniformity and stability as shall be specified under the regulations. (3) For the purposes of sub-sections (1) and (2), as the case may be, a new variety shall be deemed to be: (*a*) novel, if, at the date of filing of the application for registration for protection, the propagating or harvested material of such variety has not been sold or otherwise disposed of by or with the consent of its breeder or his successor for the purposes of exploitation of such variety (*i*) in India, earlier than one year; or (*ii*) outside India, in the case of trees or vines earlier than six years, or in any other case, earlier than four years, before the date of filing such application.

(b) distinct, if it is clearly distinguishable by at least one essential characteristic from any another variety whose existence is a matter of common knowledge in any country at the time of filing of the application.

(c) uniform, if subject to the variation that may be expected from the particular features of its propagation it is sufficiently uniform in its essential characteristics;

(d) stable, if its essential characteristics remain unchanged after repeated propagation or, in the case a particular cycle of propagation, at the end of each such cycle.

(4) A new variety shall not be registered under this Act if the denomination given to such variety: (*i*) is not capable of identifying such variety; or (*ii*) consists solely of figures; or (*iii*) is liable to mislead or to cause confusion concerning the characteristics, value identity of such variety or the identity of breeder of such variety; or (*iv*) is not different from every denomination which designates a variety of the same botanical species or of a closely related species registered under this Act; or (*v*) is likely to deceive the public or cause

confusion in the public regarding the identity of such variety; or (*vi*) is likely to hurt the religious sentiments respectively of any class or section of the citizens of India; or (*vii*) is prohibited for use as a name or emblem for any of the purposes mentioned in section 3 of the Emblems and Names (Prevention of Improper Use) Act, 1950 (12 of 1950); or (*viii*) is comprised of solely or partly of geographical name.

16. Persons Who May Make Application: An application for registration under section 14 shall be made by: (*a*) any person claiming to be the breeder of the variety; or (*b*) any successor of the breeder of the variety; or (*c*) any person being the assignee of the breeder of the variety in respect of the right to make such application; or (*d*) any farmer or group of farmers or community of farmers claiming to be the breeder of the variety; or (*e*) any person authorized in the prescribed manner by a person specified under clauses (*a*) to (*d*) to make application on his behalf; or (*f*) any university or publicly funded agricultural institution claiming to be the breeder of the variety.

17. Compulsory Variety Denomination: Every application shall assign a single and distinct denomination to a variety with respect to which he is seeking registration under this Act in accordance with the regulations.

18. Form of Application: (1) Every application for registration under section 14 shall— (*a*) be with respect to a variety; (*b*) state the denomination assigned to such variety by the applicant; (*c*) be accompanied by an affidavit sworn by the applicant that such variety does not contain any gene or gene sequence involving terminator technology; (*d*) be in such form as may be specified by regulations; (*e*) contain a complete passport data of the parental lines from which the variety has been derived along with the geographical location in India from where the genetic material has been taken and all such information relating to the contribution, if any, of any farmer, village community, institution or organization in breeding, evolving or developing the variety; (*f*) be accompanied by a statement containing a brief description of the variety bringing out its characteristics of novelty, distinctiveness, uniformity and stability as required for registration; (*g*) be accompanied by such fees as may be prescribed; (*h*) contain a declaration that the genetic material or parental material acquired for breeding, evolving or developing the variety has been lawfully acquired.

Provided that in case where the application is for the registration of farmers' variety, nothing contained in clauses (*b*) to (*i*) shall apply in respect of the application and the application shall be in such form as may be prescribed.

(2) Every application referred to in sub-section (*1*) shall be filed in the office of the Registrar.

(3) Where such application is made by virtue of a succession or an assignment of the right to apply for registration, there shall be furnished at the time of making the application, or within such period after making the application as may be prescribed, a proof of the right to make the application.

19. Test to Be Conducted: (1) Every applicant shall, along with the application for registration made under this Act, make available to the Registrar such quantity of seed of a variety for registration of which such application is made, for the purpose of conducting tests to evaluate whether seed of such variety along with parental material conform to the standards as may be specified by regulations. (2) The applicant shall deposit such fees as may be prescribed for conducting tests referred to in sub-section (1). (3) The tests referred to in sub-section (*1*) shall be conducted in such manner and by such method as may be prescribed.

20. Acceptance of Application or Amendment Thereof: (1) On receipt of an application under section 14, the Registrar may, after making such inquiry as he thinks fit with respect to the particulars contained in such application, accept the application absolutely or subject to such conditions or limitations as he deems fit. (2) Where the Registrar is satisfied that the application does not comply with the requirements of this Act or any rules or regulations made thereunder, he may, either: (*a*) require the applicant to amend the application to his satisfaction; or (*b*) reject the application:

21. Advertisement of Application

(1) Where an application for registration of a variety has been accepted absolutely or subject to conditions or limitations under sub-section (1) of section 20, the Registrar shall, as soon as after its acceptance, cause such application together with the conditions or limitations, if any, subject to which it has been accepted and the specifications of the variety for registration of which such application is made including its photographs or drawings, to be advertised in the prescribed manner calling objections from the persons interested in the matter.

(2) Opposition to the registration under sub-section (*2*) may be made on any of the following grounds, namely: (*a*) that the person opposing the application is entitled to the breeder's right as against the applicant; or (*b*) that the variety is not registrable under this Act; or (*c*) that the grant of certificate of registration may not be in public interest; or (*d*) that the variety may have adverse effect on the environment.

(3) The Registrar shall serve a copy of the notice of opposition on the applicant for registration and, within two months from the receipt by the applicant of such copy of the notice of opposition, the applicant shall send to the Registrar in the prescribed manner a counter-statement of the grounds on which he relies for his application, and if he does not do so, he shall be deemed to have abandoned his application.

(4) If the applicant sends such counter-statement, the Registrar shall serve a copy thereof on the person giving notice of opposition.

(5) Any evidence upon which the opponent and the applicant may rely shall be submitted, in the manner prescribed and within the time prescribed, to the Registrar and the Registrar shall give an opportunity to them to be heard, if so desired.

(6) The Registrar shall, after hearing the parties, if so required, and considering the evidence, decide whether and subject to what conditions or limitations, if any, the registration is to be permitted and may take into account a ground of objection whether relied upon by the opponent or not.

(7) Where a person giving notice of opposition or an applicant sending a counter-statement after receipt of a copy of such notice neither resides nor carries on business in India, the Registrar may require him to give security for the cost of proceedings before him and in default of such security being duly given may treat the opposition or application, as the case may be, as abandoned.

(8) The Registrar may, on request, permit correction of any error in, or any amendment of, a notice of opposition or a counter-statement on such terms as he may think fit.

22. Registrar to Consider Grounds for Opposition: The Registrar shall consider all the grounds on which the application has been opposed and after giving reasons for his decision, by order, uphold or reject the opposition.

23. Registration of Essentially Derived Variety

(1) An application for the registration of an essentially derived variety of the genera or species specified under sub-section (2) of section 29 by the Central Government shall be made to the Registrar by or on behalf of any person referred to in section 14 and in the manner specified in section 18 as if for the word “variety”, the words “essentially derived variety” have been substituted therein and shall be accompanied by such documents and fees as may be prescribed.

(2) When the Registrar is satisfied that the requirements of sub-section (1) have been complied with to his satisfaction, he shall forward the application with his report and the entire relevant document to the Authority.

(3) On receipt of an application under sub-section (2), the Authority shall get examined such essentially derived variety to determine as to whether the essentially derived variety is a variety derived from the initial variety by conducting such tests and following such procedure as may be prescribed.

(4) When the Authority is satisfied on the report of the test referred to in sub-section (*3*) that the essentially derived variety has been derived from the initial variety, it may direct the Registrar to register such essentially derived variety and the Registrar shall comply with the direction of the Authority.

(5) Where the Authority is not satisfied on the report of the test referred to in sub-section (*3*) that the essentially derived variety has been derived from the initial variety it shall refuse the application.

(6) The rights of the breeder of a variety contained in section 28 shall apply to the breeder of essentially derived variety.

(7) An essentially derived variety shall not be registered under this section unless it satisfies the requirements of section 15 as if for the word "variety", the words "essentially derived variety" have been substituted therein.

(8) When an essentially derived variety has been registered by the Registrar in compliance with the direction of the Authority under sub-section (4), the Registrar shall issue to the applicant a certificate of registration in the prescribed form and sealed with seal of the Registry and send a copy thereof to the Authority and to such other authority, as may be prescribed, for information.

24. Issue of Certificate of Registration

(1) When an application for registration of a variety (other than an essentially derived variety) has been accepted and either: (*a*) the application has not been opposed and the time of notice of opposition has expired; or (*b*) the application has been opposed and opposition has been rejected, the Registrar shall register the variety.

(2) On the registration of the variety (other than an essentially derived variety), the Registrar shall issue to the applicant a certificate of registration in the prescribed form and sealed with the seal of the Registry and send a copy to the Authority for determination of benefit sharing and to such other authority, as may be prescribed, for information.

(3) Where registration of a variety (other than an essentially derived variety), is not completed within twelve months from the date of the application by reason of default on the part of the applicant, the Registrar may, after giving notice to the applicant in the prescribed manner, treat the application as abandoned unless it is completed within the time specified in that behalf in the notice.

(4) The Registrar may amend the register or a certificate of registration for the purpose of correcting a clerical error or an obvious mistake.

(5) The Registrar shall have power to issue such directions to protect the interests of a breeder against any abusive act committed by any third party during the period between filing of application for registration and decision taken by the Authority on such application.

(6) The certificate of registration issued under this section or sub-section (8) of section 23 shall be valid for nine years in the case of trees and vines and six years in the case of other crops and may be reviewed and renewed for remaining period on payment of such fees as may be fixed by the rules made in this behalf subject to the condition that the total period of validity shall not exceed:

(*i*) in the case of trees and vines, eighteen years from the date of registration of the variety; (*ii*) in the case of extant variety, fifteen years from the date of the notification of that variety by the Central Government under section 5 of the Seeds Act, 1966 (54 of 1966); and (*iii*) in other cases, fifteen years from the date of registration of the variety.

25. Publication of List of Varieties: The Authority shall, within such intervals as it thinks appropriate, publish the list of varieties which have been registered during that interval.

26. Determination of Benefit Sharing by Authority

(1) On receipt of a copy of the certificate of registration under sub-section (8) of section 23 or sub-section (2) of section 24, the Authority shall publish such contents of the certificate and invite claims of benefit sharing to the variety registered under such certificate in the manner as may be prescribed.

(2) On invitation of the claims under sub-section (*1*), any person or group of persons or firm or governmental or non-governmental organization shall submit its claim of benefit sharing to such variety in the prescribed form within such period, and accompanied with such fees, as may be prescribed: Provided that such claim shall only be submitted by any (*i*) person or group of persons, if such person or every person constituting such group is a citizen of India; or (*ii*) firm or governmental or non-governmental organization, if such firm or organization is formed or established in India.

(3) On receiving a claim under sub-section (*2*), the Authority shall send a copy of such claim to the breeder of the variety registered under such certificate and the breeder may, on receipt of such copy, submit his opposition to such claim within such period and in such manner as maybe prescribed.

(4) The Authority shall, after giving an opportunity of being heard to the parties, dispose of the claim received under sub-section (*2*).

(5) While disposing of the claim under sub-section (*4*), the Authority shall explicitly indicate in its order the amount of the benefit sharing, if any, for which the claimant shall be entitled and shall take into consideration the following matters, namely: (*a*) the extent and nature of the use of genetic material of the claimant in the development of the variety relating to which the benefit sharing has been claimed; (*b*) the commercial utility and demand in the market of the variety relating to which the benefit sharing has been claimed.

(6) The amount of benefit sharing to a variety determined under this section shall be deposited by the breeder of such variety in the manner referred to in clause (*a*) of sub-section (1) of section 45 in the National Gene Fund.

(7) The amount of benefit sharing determined under this section shall, on a reference made by the Authority in the prescribed manner, be recoverable as an arrear of land revenue by the District Magistrate within whose local limits of jurisdiction the breeder liable for such benefit sharing resides.

27. Breeder to Deposit Seeds or Propagating Material: The breeder shall be required to deposit such quantity of seeds or propagating material including parental line seeds of registered variety in the National Gene Bank as may be specified in the regulations for reproduction purpose at the breeder's expense within such time as may be specified in that regulation.

28. Registration to Confer Right

(1) Subject to the other provisions of this Act, a certificate of registration for a variety issued under this Act shall confer an exclusive right on the breeder or his successor, his agent or licensee, to produce, sell, market, distribute, import or export the variety.

(2) A breeder may authorize any person to produce, sell, market or otherwise deal with the variety registered under this Act subject to such limitations and conditions as may be specified by regulations.

(3) Every authorization under this section shall be in such form as may be specified by regulations.

(4) Where an agent or a licensee referred to in sub-section (1) becomes entitled to produce, sell, market, distribute, import or export a variety, he shall apply in the prescribed manner and with the prescribed fees to the Registrar to register his title and the registrar shall, on receipt of application and on proof of title to his satisfaction, register him as an agent or a licensee, as the case may be, in respect of the variety for which he is entitled for such right, and shall cause particulars of such entitlement and conditions or restrictions, if any, subject to which such entitlement is made, to be entered in the Register.

(5) The Registrar shall issue a certificate of registration under sub-section (4) to the application after such registration and shall enter in the certificate the brief conditions of entitlement, if any, in the prescribed manner, and such certificate shall be the conclusive proof of such entitlement and the conditions or restriction thereof, if any.

(6) Subject to any agreement subsisting between the parties, an agent or licensee of a right to a variety registered under sub-section (4) shall be entitled to call upon the breeder or his successor thereof to take proceedings to prevent infringement thereof, and if the breeder or his successor refuses or neglects to do so within three months after being so called upon, such registered agent or licensee may institute proceedings for infringement in his own name as if he were the breeder, making the breeder or his successor a defendant.

(7) Notwithstanding anything contained in any other law, a breeder or his successor so added as defendant shall not be liable for any costs unless he enters an appearance and takes part in the proceedings.

(8) Nothing in this section shall confer on a registered agent or registered licensee of a variety any right to transfer such right further thereof.

(9) Without prejudice to the registration under sub-section (4), the terms of registration—

(*a*) may be varied by the Registrar as regards the variety in respect of which, or any condition or restriction subject to which, it has effect on receipt of an application in the prescribed manner of the registered breeder of such variety or his successor; (*b*) may be cancelled by the Registrar on the application in the prescribed manner of the registered breeder of such variety or his successor or of the registered agent or registered licensee of such variety; (*c*) may be cancelled by the Registrar on the application in the prescribed manner of any person other than the breeder, his successor, the registered agent or the registered licensee on any of the following grounds, namely, (*i*) that the breeder of a variety or his successor or the registered agent or registered licensee of such variety, misrepresented, or failed to disclose, some fact material to the application for registration under sub-section (4) which if accurately represented or disclosed would have justified the refusal of the application for registration of the registered agent or registered licensee; (*ii*) that the registration ought not to have effected having regard to the right vested in the applicant by virtue of a contract in the performance of which he is interested; (*d*) may be cancelled by the Registrar on the application in the prescribed manner of the breeder of a registered variety or his successor on the ground that any stipulation in the agreement between the registered agent or the registered licensee, as the case may be, and such breeder or his successor regarding the variety for which such agent or licensee is registered is not being enforced or is not being complied with; (*e*) may be cancelled by the registrar on the application of any person in the prescribed manner on the ground that the variety relating to the registration is no longer existing.

(10) The registrar shall issue notice in the prescribed manner of' every application under this section to the registered breeder of a variety or his successor and to each registered agent or registered licensee (not being the applicant) of such variety.

(11) The Registrar shall, before making any order under sub-section (9) forward the application made in that behalf along with any objection received by any party after notice under sub-section (10) for the consideration of the Authority, and the Authority may, after making such inquiry as it thinks fit, issue such directions to the Registrar as it thinks fit and the Registrar shall dispose of the application in accordance with such directions.

29. Exclusion of Certain Varieties

(1) Notwithstanding anything contained in this Act, no registration of a variety shall be made under this Act in cases where prevention of commercial exploitation of such variety is necessary to protect public order or public morality or human, animal and plant life and health or to avoid serious prejudice to the environment, including genetic use restriction technology and terminator technology.

(2) The Central Government shall, by notification in the Official Gazette, specify the genera or species for the purposes of registration of varieties other than extant varieties and farmers' varieties under this Act.

30. Researcher's Rights: Nothing contained in this Act shall prevent:

(a) the use of any variety registered under this Act by any person using such variety for conducting experiment or research; or (*b*) the use of a variety by any person as an initial source of variety for the purpose of creating other varieties: Provided that the authorization of the breeder of a registered variety is required where the repeated use of such variety as a parental line is necessary for commercial production of such other newly developed variety.

Clause No. 31 to 38 are omitted from this write-up as these mostly belong its applicability to registration from citizens of convention countries, provision of reciprocity, surrender of certificate of registration, revocation of protection on certain grounds, payment of annual fees, power to cancel registration, correction of register, and alternation of denomination of a registered variety. These are not core issues related to registration of a variety and interested readers may refer to Ram (2012; 2019) and the website https://indiacode.nic.in PDF

39. Farmer's Rights

(1) Notwithstanding anything contained in this Act: (*i*) a farmer who has bred or developed a new variety shall be entitled for registration and other protection in like manner as a breeder of a variety under this Act; (*ii*) the farmers' variety shall be entitled for registration if the application contains declarations as specified in clause (*h*) of sub-section (1) of section 18; (*iii*) a farmer who is engaged in the conservation of genetic resources of land races and wild relatives of economic plants and their improvement through selection and preservation shall be entitled in the prescribed manner for recognition and reward from the Gene Fund: Provided that material so selected and preserved has been used as donors

of genes in varieties registrable under this Act; (*iv*) a farmer shall be deemed to be entitled to save, use, sow re-sow, exchange, share or sell his farm produce including seed of a variety protected under this Act in the same manner as he was entitled before the coming into force of this Act: Provided that the farmer shall not be entitled to sell branded seed of a variety protected under this Act.

(2) Where any propagating material of a variety registered under this Act has been sold to a farmer or a group of farmers or any organization of farmers, the breeder of such variety shall disclose to the farmer or the group of farmers or the organization of farmers, as the case may be, the expected performance under given conditions, and if such propagating material fails to provide such performance under such given conditions, the farmer or the group of farmers or the organization of farmers, the case may be, may claim compensation in the prescribed manner before the Authority, and Authority may take further follow up action.

40. Certain Information to be Given in Application for Registration

(1) A breeder or other person making application for registration of any variety under Chapter III shall disclose in the application the information regarding the use of genetic material conserved by any tribal or rural families in the breeding or development of such variety failing which the application may be rejected.

Rights of Communities

(1) Any person or group of persons (whether actively engaged in farming or not) or any governmental or non-governmental organization may, on behalf of any village or local community in India, file in any centre notified, with the previous approval of the Central Government, by the Authority, in the Official Gazette, any claim attributable to the contribution of the people of that village or local community, as the case may be, in the evolution of any variety for the purpose of staking a claim on behalf of such village or local community.

(2) Where any claim is made under sub-section (1), the centre notified under that sub-section may verify the claim made by such person or group of persons or such governmental or non-governmental organization in such manner as it deems fit, and if it is satisfied that such village or local community has contributed significantly to the evolution of the variety which has been registered under this Act, it shall report its findings to the Authority.

(3) When the Authority, on a report under sub-section (2) is satisfied, after such inquiry as it may deem fit, that the variety with which the report is related has been registered under the provisions of this Act, it may issue notice in the prescribed manner to the breeder of that variety and after providing opportunity to such breeder to file objection in the prescribed manner and of being heard, it may subject to any limit notified by the Central Government, by order, grant such sum of compensation to be paid to a person or group of persons or governmental or non-governmental organization which has made claim under sub-section (1*l*), as it may deem fit.

(4) Any compensation granted under sub-section (3) shall be deposited by the breeder of the variety in the Gene Fund.

42. Protection of Innocent Infringement: Notwithstanding anything contained in this Act: (*i*) a right established under this Act shall not be deemed to be infringed by a farmer who at the time of such infringement was not aware of the existence of such right; and (*ii*) a relief which a court may grant in any suit for infringement referred to in section 65 shall not be granted by such court, nor any cognizance of any offence under this Act shall be taken, for such infringement by any court against a farmer who proves, before such court, that at the time of the infringement he was not aware of the existence of the right so infringed.

43. Authorization of Farmers' Variety: Notwithstanding anything contained in sub-section (6) of section 23 and section 28, where an essentially derived variety is derived from a farmers' variety, the authorization under sub-section (2) of section 28 shall not be given by the breeder of such farmers' variety except with the consent of the farmers or group of farmers or community of farmers who have made contribution in the preservation or development of such variety.

44. Exemption From Fees: A farmer or group of farmers or village community shall not be liable to pay any fees in any proceeding before the Authority or Registrar or the Tribunal or the High Court under this Act or the rules made thereunder.

45. Gene Fund: (1) The Central Government shall constitute a Fund to be called the National Gene Fund and there shall be credited thereto: (*a*) the benefit sharing received in the prescribed manner from the breeder of a variety or an essentially derived variety registered under this Act, or propagating material of such variety or essentially derived variety, as the case may be; (*b*) the annual fees payable to the Authority by way of royalty under sub-section (1) of section 35; (*c*) the compensation deposited in the Gene Fund under sub-section (4) of section 41; (*d*) the contribution from any national and international organization and other sources. (2) The Gene Fund shall, in the prescribed manner, be applied for meeting: (*a*) any amount to be paid by way of benefit sharing under sub-section (5) of section 26; (*b*) the compensation payable under sub-section (3) of section 41; (*c*) the expenditure for supporting the conservation and sustainable use of genetic resources including *in-situ* and *ex-situ* collections and for strengthening the capability of the Panchayat in carrying out such conservation and sustainable use; (*d*) the expenditures of the schemes relating to benefit sharing framed under section 46.

46. Framing of Schemes, etc: (1) The Central Government shall, for the purposes of section 41 and clause (*d*) of sub-section (2) of section 45, frame, by notification in the Official Gazette, one or more schemes. (2) In particular, and without prejudice to the generality of the provisions of sub-section (1), the scheme may provide for all or any of the following matters, namely: (*a*) the registration of the claims for the purposes of section 41 under the scheme and all matters connected with such registration; (*b*) the processing of such claims for securing their enforcement and matters connected therewith; (*c*) the maintenance of records and Registers in respect of such claims; (*d*) the utilisation, by way of disbursal (including apportionment) or otherwise, of any amounts received in satisfaction of such claims; (*e*) the procedure for disbursal or apportionment by the Authority in the event of dispute regarding such claims; (*f*) the utilisation of benefit sharing for the purposes relating to breeding, discovery or development of varieties; (*g*) the maintenance and audit of accounts with respect to the amounts referred to in clause (*d*).

47. Power of Authority to Make Order for Compulsory License in Certain Circumstances

(1) At any time, after the expiry of three years from the date of issue of a certificate of registration of a variety, any person interested may make an application to the Authority alleging that the reasonable requirements of the public for seed or other propagating material of the variety have not been satisfied or that the seed or other propagating material of the variety is not available to the public at a reasonable price and pray for the grant of a compulsory license to undertake production, distribution and sale of the seed or other propagating material of that variety.

(2) Every application under sub-section (1) shall contain a statement of the nature of the applicant's interest together with such particulars as may be prescribed.

(3) The Authority, after consultation with Central Government, and if satisfied, may order such breeder to grant a license to the applicant upon such terms and conditions as it may deem fit.

48. When Requirement of Public Deemed to Have Not Been Satisfied: In determining the question as to whether the reasonable requirements of the public for seeds of a variety or its propagating material as referred to in sub-section (1) or sub-section (3) of section 47, the Authority shall take into account- (*i*) the nature of the variety, the time which has elapsed since the grant of the certificate of registration of the variety, price of the seed of the variety and the measures taken by the breeder or any registered licensee of the variety to meet the requirement of the public; and (*ii*) the capacity, ability and technical competence of the applicant to produce and market the variety to meet the requirement of the public.

49. Adjournment of Application for Grant of Compulsory License: (1) If the breeder of a variety registered under this Act in respect of which any application has been pending before the Authority under section 47 makes a written request to the Authority on the ground that due to any reasonable factor, such breeder has been unable to produce seed or other propagating material of the variety on a commercial scale to an adequate extent till the date of making such request, the Authority may, on being satisfied that the said ground is reasonable, adjourn the hearing of such application for such period not exceeding twelve months in aggregate as it may consider sufficient for optimum production of the seed or propagating material of such variety, as the case may be, by such breeder.

50. Duration of Compulsory License: The Authority shall determine the duration of the compulsory licenses granted under this Chapter and such duration may vary from case to case keeping in view the gestation periods and other relevant factors but in any case it shall not exceed the total remaining period of the protection of the variety.

(i) **Authority to Settle Terms and Conditions of Licenser:** (1) The Authority shall, while determining the terms and conditions of a compulsory license under the provisions of this Chapter, endeavor to secure: reasonable compensation to the breeder of the variety relating to the compulsory license;

(ii) that the compulsory licensee of such variety possesses the adequate means to provide to the farmers, the seeds or its other propagating material of such variety timely and at reasonable market price.

(2) No compulsory license granted by the Authority shall authorize the licensee to import the variety relating to such license.

52. Revocation of Compulsory Licenser: (1) The Authority, may on its own motion or on application from an aggrieved person made to it in the prescribed form, if it is satisfied that a compulsory licensee registered under this Chapter has violated any terms or conditions of his license or it is not appropriate to continue further such license in public interest, it may make order to revoke such license.

(2) When a license is revoked under sub-section (1) by an order of the Authority, the Authority shall send a copy of such order to the Registrar to rectify the entry or correct the Register relating to such revocation.

53. Modification of Compulsory License: The Authority may, on its own motion or on application from licensee of a compulsory license, if it considers, in public interest, so to do, modify, by order, such terms and conditions as it thinks fit and send a copy of such order to the Registrar to correct the entries in the Register.

The matters under clauses 54 to 97 dealing with tribunals, appeals, filing suits, budget of Authority, financial and administrative powers of the Chairman, power of central government to make rules, etc are not included here as these do not directly relate to the procedures of registration of crop varieties and the related matters. However, interested readers are advised to refer to Ram (2012; 2019) and https://indiacode.nic.in PDF for details, if needed.

7

Biological Diversity Act-2002

The increasing concerns about dwindling biological resources globally led to the Convention on Biological Diversity in 1992. This convention for the first time, recognized sovereign rights of States over their biological resources and emphasized that access to genetic resources should be only for environmentally sound purposes and should be subject to national legislations. The access has to be on mutually agreed terms which inter-alia would include recognition of associated Traditional Knowledge of indigenous communities and equitable benefit sharing arrangements.

Taking cognizance of the provisions of the CBD, and to deal with extensive pressure on our biological resources, India has enacted an umbrella legislation called Biological Diversity Act-2002 (No. 18 of 2003), and also notified the Biological Diversity Rules-2004 for guidance and compliance by various stakeholders, including the Union and State Governments, Non-state sectors, and individuals. The Act aims at conservation of biological resources as well as facilitating access to them in a sustainable manner and through a just process. The salient features of this act are as follows (NBA, 2004).

1. Short Title, Extent and Commencement

(1) This Act may be called the Biological Diversity Act- 2002.

(2) It extends to the whole of India.

2. Definitions

(a) "benefit claimers" means the conservers of biological resources, their byproducts, creators and holders of knowledge and information relating to the use of such biological resources, innovations and practices associated with such use and application;

(b) "biological diversity" means the variability among living organisms from all sources and the ecological complexes of which they are part, and includes diversity within species or between species and of eco-systems;

(c) "biological resources" means plants, animals and micro-organisms or parts thereof, their genetic material and by-products (excluding value added products) with actual or potential use or value, but does not include human genetic material;

(d) "bio-survey and bio-utilization" means survey or collection of species, subspecies, genes, components and extracts of biological resource for any purpose and includes characterization, inventorisation and bioassay;

(e) "Chairperson" means the Chairperson of the National Biodiversity Authority or, as the case may be, of the State Biodiversity Board;

(f) "commercial utilization" means end uses of biological resources for commercial utilization such as drugs, industrial enzymes, food flavours, fragrance, cosmetics, emulsifiers, oleoresins, colours, extracts and genes used for improving crops and livestock through genetic intervention, but does not include conventional breeding or traditional practices in use in any agriculture, horticulture, poultry, dairy farming, animal husbandry or bee keeping;

(g) "fair and equitable benefit sharing" means sharing of benefits as determined by the National Biodiversity Authority under section 2 1;

(h) "local bodies" means Panchayats and Municipalities, by whatever name called, within the meaning of clause (1) of article 243B and clause (1) of article 243Q of the Constitution and in the absence of any Panchayats or Municipalities, institutions of self-government constituted under any other provision of the Constitution or any Central Act or State Act;

(i) "member" means a member of the National Biodiversity Authority or a State Biodiversity Board and includes the Chairperson;

(j) "National Biodiversity Authority" means the National Biodiversity Authority established under section 8;

(k) "research" means study or systematic investigation of any biological resource or technological application, that uses biological systems, living organisms or derivatives thereof to make or modify products or processes for any use;

(l) "State Biodiversity Board" means the State Biodiversity Board established under section 22;

(m) "sustainable use" means the use of components of biological diversity in such manner and at such rate that does not lead to the long-term decline of the biological diversity thereby maintaining its potential to meet the needs and aspirations of present and future generations;

(n) "value added products" means products which may contain portions or extracts of plants and animals in unrecognizable and physically inseparable form.

Certain Persons Not to Undertake Biodiversity Related Activities Without Approval of National Biodiversity Authority

3.(1) No person referred to in sub-section (2) shall, without previous approval of the National Biodiversity Authority, obtain any biological resource occurring in India or knowledge associated thereto for research or for commercial utilization or for bio-survey and bio-utilization.

(2) The persons who shall be required to take the approval of the National Biodiversity Authority under sub-section (1) are the following, namely:

(a) a person who is not a citizen of India;

(b) a citizen of India, who is a non-resident as defined in clause (30) of section 2 of the Income-tax Act, 196 1;

(c) a body corporate, association or organization-

(i) not incorporated or registered in India; or

(ii) incorporated or registered in India under any law for the time being in force which has any non-Indian participation in its share capital or management.

Results of Research Not to Be Transferred to Certain Persons Without Approval of National Biodiversity Authority

4. No person shall, without the previous approval of the National Biodiversity Authority, transfer the results of any research relating to any biological resources occurring in, or obtained from, India for monetary consideration or otherwise to any person who is not a citizen of India or citizen of India who is non-resident as defined in clause (30) of section 2 of the Income-tax Act, 1961 or a body corporate or organization which is not registered or incorporated in India or which has any non-Indian participation in its share capital or management.

Explanation.- For the purposes of this section, "transfer" does not include publication of research papers or dissemination of knowledge in any seminar or workshop, if such publication is as per the guidelines issued by the Central Government.

Sections 3 and 4 Not to Apply to Certain Collaborative Research Projects

5.(1) The provisions of sections 3 and 4 shall not apply to collaborative research projects involving transfer or exchange of biological resources or information relating thereto between institutions, including Government sponsored institutions of India, and such institutions in other countries, if such collaborative research projects satisfy the conditions specified in sub-section (3).

(2) All collaborative research projects, other than those referred to in sub-section (1) which are based on agreements concluded before the commencement of this Act and in force shall, to the extent the provisions of agreement are inconsistent with the provisions of this Act or any guidelines issued under clause (a) of sub-section (3), be void.

(3) For the purposes of sub-section (1), collaborative research projects shall-

(a) conform to the policy guidelines issued by the Central Government in this behalf;

(b) be approved by the Central Government.

Application for Intellectual Property Rights Not to Be Made Without Approval of National Biodiversity Authority

6.(1) No person shall apply for any intellectual property right, by whatever name called, in or outside India for any invention based on any research or information on a biological resource obtained from India without obtaining the previous approval of the National Biodiversity Authority before making such application.

Provided that if a person applies for a patent, permission of the National Biodiversity Authority may be obtained after the acceptance of the patent but before the seating of tile patent by the patent authority concerned:

(2) The National Biodiversity Authority may, while granting the approval under this section, impose benefit sharing fee or royalty or both or impose conditions including the sharing of financial benefits arising out of the commercial utilization of such rights.

(3) The provisions of this section shall not apply to any person making an application for any right under any law relating to protection of plant varieties enacted by Parliament.

(4) Where any right is granted under law referred to in sub-section (3), the concerned authority granting such right shall endorse a copy of such document granting the right to the National Biodiversity Authority.

Prior Intimation to State Biodiversity Board for Obtaining Biological Resource for Certain Purposes

7. No person, who is a citizen of India or a body corporate, association or organization which is registered in India, shall obtain any biological resource for commercial utilization, or bio-survey and bio-utilization for commercial utilization except after giving prior intimation to the State Biodiversity Board concerned:

provided that the provisions of this section shall not apply to the local people and communities of the area, including growers and cultivators of biodiversity, and *vaids* and *hakims*, who have been practicing indigenous medicine.

Establishment of National Biodiversity Authority

8.(1) With effect from such date as the Central Government may, appoint, there shall be established by the Central Government for the purposes of this Act, a body to be called the National Biodiversity Authority.

(2) The National Biodiversity Authority shall be a body corporate by the name aforesaid, having perpetual succession and a common seal, with power to acquire, hold and dispose of property, both movable and immovable, and to contract, and shall by the said name sue and be sued.

(3) The head office of the National Biodiversity Authority shall be at Chennai and the National Biodiversity Authority may, with the previous approval of the Central Government, establish offices at other places in India.

(4) The National Biodiversity Authority shall consist of a Chairperson and Members in accordance with the provisions of the Act.

Conditions of Service of Chairperson and Members/Chairperson to Be Chief Executive of NBA, Removal of Members, Meetings of NB

Covered under clause 9-12 not included here as these are purely administrative matters.

Committees of National Biodiversity Authority

13.(1)The National Biodiversity Authority may constitute a committee to deal with agro-biodiversity.

(2) Without prejudice to the provisions of sub-section (1), the National Biodiversity Authority may constitute such number of committees as it deems fit for the efficient discharge of its duties and performance of its functions under this Act.

(3) A committee constituted under this section shall co-opt such number of persons, who are not the members of the National Biodiversity Authority, as it may think fit and the persons so co-opted shall have the right to attend the meetings of the committee and take part in its proceedings but shall not have the right to vote.

(4) The persons appointed as members of the committee under sub-section (2) shall be entitled to receive such allowances or fees for attending the meetings of the committee as may be fixed by the Central Government.

Officers and Employees of National Biodiversity Authority/ Authentication of Orders and Decisions of National Biodiversity Authority/ Delegation of Powers/ Expenses of National Biodiversity Authority to Be Defrayed Out of the Consolidated Fund of India

These items covered under clauses 14-17 are excluded here as these are purely administrative matters.

Functions and Powers of National Biodiversity Authority

18.(1)It shall be the duty of the National Biodiversity Authority to regulate activities referred to in sections 3, 4 and 6 and by regulations issue guidelines for access to biological resources and for fair and equitable benefit sharing.

(2) The National Biodiversity Authority may grant approval for undertaking any activity referred to in sections 3, 4 and 6.

(3) The National Biodiversity Authority may-

(a) advise the Central Government on matters relating to the conservation of biodiversity, sustainable use of its components and equitable sharing of benefits arising out of the utilization of biological resources;

(b) advise the State Governments in the selection of areas of biodiversity importance to be notified under sub-section (1) of section 37 as heritage sites and measures for the management of such heritage sites;

(c) perform such other functions as may be necessary to carry out the provisions of this Act.

(4) The National Biodiversity Authority may, on behalf of the Central Government, take any measures necessary to oppose the grant of intellectual property rights in any country outside India on any biological resource obtained from India or knowledge associated with such biological resource which is derived from India.

Approval by National Biodiversity Authority for Undertaking Certain Activities

19.(1) Any person referred. to in sub-section (2) of section 3 who intends to obtain any biological resource occurring in India or knowledge associated thereto for research or for commercial utilization or for bio-survey and bio-utilization or transfer the results of any research relating to biological resources occurring in, or obtained from, India, shall make application in such form and payment of such fees as may be prescribed, to the National Biodiversity Authority.

(2) Any person who intends to apply for a patent or any other form of intellectual property protection whether in India or outside India referred to in sub-section (1) of section 6, may make an application in such form and in such manner as may be prescribed to the National Biodiversity Authority.

(3) On receipt of an application under sub-section (1) or sub-section (2), the National Biodiversity Authority may, after making such enquiries as it may deem fit and if necessary after consulting an expert committee constituted for this purpose, by order, grant approval subject to any regulations made in this behalf and subject to such terms and conditions as it may deem fit, including the imposition of charges by way of royalty or for reasons to be recorded in writing, reject the application.

(4) The National Biodiversity Authority shall give public notice of every approval granted by it under this section.

Transfer of Biological Resource or Knowledge

20.(1) No person who has been granted approval under section 19 shall transfer any biological resource or knowledge associated thereto which is the subject matter of the said approval except with the permission of the National Biodiversity Authority.

(2) Any person who intends to transfer any biological resource or knowledge associated thereto referred to in sub-section (1) shall make an application in such form and in such manner as may be prescribed to the National Biodiversity Authority.

(3) On receipt of an application under sub-section (2), the National Biodiversity Authority may, after making such enquiries as it may deem fit and if necessary after consulting an expert committee constituted for this purpose, by order, grant approval subject to such terms and conditions as it may deem fit, including the imposition of charges by way of royalty or for reasons to be recorded in writing, reject the application:

(4) The National Biodiversity Authority shall give public notice of every approval granted by it under this section.

Determination of Equitable Benefit Sharing by National Biodiversity Authority

21.(1)The National Biodiversity Authority shall while granting approvals under section 19 or section 20 ensure that the terms and conditions subject to which approval is granted secures equitable sharing of benefits arising out of the use of accessed biological resources, their by-products, innovations and practices associated with their use and applications and knowledge relating thereto in accordance with mutually agreed terms and conditions between the person applying for such approval, local bodies concerned and the benefit claimers.

(2) The National Biodiversity Authority shall, subject to any regulations made in this behalf, determine the benefit sharing which shall be given effect in all or any of the following manner, namely:

(a) grant of joint ownership of intellectual property rights to the National Biodiversity Authority, or where benefit claimers are identified, to such benefit claimers;

(b) transfer of technology;

(c) location of production, research and development units in such areas which will facilitate better living standards to the benefit claimers;

(d) association of Indian scientists, benefit claimers and the local people with research and development in biological resources and bio-survey and bio-utilization;

(e) setting up of venture capital fund for aiding the cause of benefit claimers;

(f) payment of monetary compensation and non-monetary benefits to the benefit claimers as the National Biodiversity Authority may deem fit.

(3) Where any amount of money is ordered by way of benefit sharing, the National Biodiversity Authority may direct the amount to be deposited in the National Biodiversity Fund.

Provided that where biological resource or knowledge was a result of access from specific individual or group of individuals or organizations, the National Biodiversity Authority may direct that the amount shall be paid directly to such individual or group of individuals or organizations in accordance with the terms of any agreement and in such manner as it deems fit.

(4) For the purposes of this section, the National Biodiversity Authority shall, in consultation with the Central Government, by regulations, frame guidelines.

Establishment of State Biodiversity Board

22.(1) With effect from such date as the State Government may, by notification in the Official Gazette, appoint in this behalf, there shall be established by that Government for the purposes of this Act, a Board for the State to be known as the (name of the State) Biodiversity Board.

(2) In relation to a Union territory, the National Biodiversity Authority shall exercise the powers and perform the functions of a State Biodiversity Board for that Union territory.

(3) The Board shall consist of the Chairperson and the members in accordance with the provisions of BDA-2002. For details, NBA (2004) may be referred.

Functions of State Biodiversity Board

23. The functions of the State Biodiversity Board shall be to–

(a) advise the State Government on matters relating to the conservation of biodiversity, sustainable use of its components and equitable sharing of the benefits arising out of the utilization of biological resources;

(b) regulate by granting of approvals or otherwise requests for commercial utilization or bio-survey and bio-utilization of any biological resource by Indians;

(c) perform such other functions as may be necessary to carry out the provisions of this Act or as may be prescribed by the State Government.

Power of State Biodiversity Board to Restrict Certain Activities

24.(1) Any citizen of India or a body corporate, organization or association registered in India intending to undertake any activity referred to in section 7 shall give prior intimation in such form as may be prescribed by the State Government to the State Biodiversity Board.

(2) On receipt of an intimation under sub-section (1), the State Biodiversity Board may prohibit or restrict any such activity if it is of opinion that such activity is detrimental or contrary to the objectives of conservation and sustainable use of biodiversity or equitable sharing of benefits arising out of such activity.

(3) Any information given in the form referred to in sub-section (1) for prior intimation shall be kept confidential and shall not be disclosed, either intentionally or unintentionally, to any person not concerned thereto.

Provisions of Sections 9 to 17 to Apply with Modifications to State Biodiversity Board

25. The provisions of sections 9 to 17 shall apply to a State Biodiversity Board and shall have effect subject to some modifications, if necessary.

Grants or Loans by the Central Government

26. The Central Government may, after due appropriation made by Parliament by law in this behalf, pay to the National Biodiversity Authority by way of grants or loans such sums of money as the Central Government may think fit for being utilized for the purposes of this Act.

Constitution of National Biodiversity Fund

27.(1) There shall be constituted a Fund to be called the National Biodiversity Fund and there shall be credited thereto --

(a) any grants and loans made to the National Biodiversity Authority under section 26;

(b) all charges and royalties received by the National Biodiversity Authority under this Act; and

(c) all sums received by the National Biodiversity Authority from such other sources as may be decided upon by the Central Government.

(2) The Fund shall be applied for–

(a) channeling benefits to the benefit claimers;

(b) conservation and promotion of biological resources and development of areas from where such biological resources or knowledge associated thereto has been accessed;

(c) socio-economic development of areas referred to in clause (b) in consultation with the local bodies concerned.

Annual report of National Biodiversity Authority

28. The National Biodiversity Authority shall prepare its annual report, giving a full account of its activities during the previous financial year and furnish, to the Central Government, before such date as may be prescribed, its audited copy of accounts together with auditor's report thereon.

Budget Accounts and Audit

These are covered under clause 29 (NBA, 2004) and are purely administrative issues.

Annual Report to Be Laid Before Parliament

30. The Central Government shall cause the annual report and auditor's report to be laid, as soon as may be after they are received, before each House of Parliament.

Grants of Money by State Government to State Biodiversity Board

31. The State Government may pay to the State Biodiversity Board by way of grants or loans such sums of money as the State Government may think fit for being utilized for the purposes of this Act.

Constitution of State Biodiversity Fund

32.(1)There shall be constituted a Fund to be called the State Biodiversity Fund and there shall be credited thereto-

(a) any grants and loans made to the State Biodiversity Board under section 31;

(b) any grants or loans made by the National Biodiversity Authority;

(c) all sums received by the State Biodiversity Board from such other sources as may be decided upon by the State Government.

(2) The State Biodiversity Fund shall be applied for-

(a) the management and conservation of heritage sites;

(b) compensating or rehabilitating any section of the people economically affected by notification under sub-section (1) of section 37;

(c) conservation and promotion of biological resources;

(d) socio-economic development of areas from where such biological resources or knowledge associated thereto has been accessed subject to any order made under section 24, in consultation with the local bodies concerned;

(e) meeting the expenses incurred for the purposes authorized by this Act.

Annual Report of State Biodiversity Board

33. The State Biodiversity Board shall prepare its annual report, giving a full account of its activities during the previous financial year, and submit a copy thereof to the State Government.

Audit of Accounts of State Biodiversity Board

34. The accounts of the State Biodiversity Board shall be maintained and audited in such manner as may, in consultation with the Accountant-General of the State, be prescribed and the State Biodiversity Board shall furnish, to the State Government, before such date as may be prescribed, its audited copy of accounts together with auditor's report thereon.

Annual Report of State Biodiversity Board to Be Laid Before State Legislature

35. The State Government shall cause the annual report and auditor's report to be laid, as soon as may be after they are received, before the House of State Legislature.

Central Government to Develop National Strategies Plans etc. for Conservation, of Biological Diversity

36.(1)The Central Government shall develop national strategies, plans, programmes for the conservation and promotion and sustainable use of biological diversity including measures for identification and monitoring of areas rich in biological resources, promotion of *in situ*, and *ex situ*, conservation of biological resources, incentives for research, training and public education to increase awareness with respect to biodiversity.

(2) Where the Central Government has reason to believe that any area rich in biological diversity, biological resources and their habitats is being threatened by overuse, abuse or neglect, it shall issue directives to the concerned State Government to take immediate ameliorative measures, offering such State Government any technical and other assistance that is possible to be provided or needed.

(3) The Central Government shall, as far as practicable wherever it deems appropriate, integrate the conservation, promotion and sustainable use of biological diversity into relevant sectoral or cross-sectoral plans, programmes and policies.

(4) The Central Government shall undertake measures,-

(i) wherever necessary, for assessment of environmental impact of that project which is likely to have adverse effect on biological diversity, with a view to avoid or minimize such effects and where appropriate provide for public participation in such assessment;

(ii) to regulate, manage or control the risks associated with the use and release of living modified organisms resulting from biotechnology likely to have adverse impact on the conservation and sustainable use of biological diversity and human health.

(5) The Central Government shall endeavour to respect and protect the knowledge of local people relating to biological diversity, as recommended by the National Biodiversity Authority through such measures, which may include registration of such knowledge at the local, State or national levels, and other measures for protection, including *sui generis* system.

Explanation

(a) "*ex situ* conservation" means the conservation of components of biological diversity outside their natural habitats;

(b) "*in situ* conservation" means the conservation of ecosystems and natural habitats and the maintenance and recovery of viable populations of species in their natural surroundings and, in the case of domesticated or cultivated species, in the surroundings where they have developed their distinctive properties.

Biodiversity Heritage Sites

37.(1) Without prejudice to any other law for the time being in force, the State Government may, from time to time in consultation with the local bodies, notify in the Official Gazette, areas of biodiversity importance as biodiversity heritage sites under this Act.

(2) The State Government, in consultation with the Central Government, may frame rules for the management and conservation of all the heritage sites.

(3) The State Government shall frame schemes for compensating or rehabilitating any person or section of people economically affected by such notification.

Power of Central Government to Notify Threatened Species

38. Without prejudice to the provisions of any other law for the time being in force, the Central Government, in consultation with the concerned State Government, may from time to time notify any species which is on the verge of extinction or likely to become extinct in the near future as a threatened species and prohibit or regulate collection thereof for any purpose and take appropriate steps to rehabilitate and preserve those species.

Power of Central Government to Designate Repositories

39.(1) The Central Government may, in consultation with the National Biodiversity Authority, designate institutions as repositories under this Act for different categories of biological resources.

(2) The repositories shall keep in safe custody the biological material including voucher specimens deposited with them.

(3) Any new taxon discovered by any person shall be notified to the repositories or any institution designated for this purpose and he shall deposit the voucher specimens with such repository or institution.

Power of Central Government to Exempt Certain Biological Resources

40. Notwithstanding anything contained in this Act, the Central Government may, in consultation with the National Biodiversity Authority, by notification in the Official Gazette, declare that the provisions of this Act shall not apply to any items, including biological resources normally traded as commodities.

Constitution of Biodiversity Management Committee

41.(1)Every local body shall constitute a Biodiversity Management Committee within its area for the purpose of promoting conservation, sustainable use and documentation of biological diversity including preservation of habitats, conservation of land races, folk varieties and cultivars, domesticated stocks and breeds of animals and microorganisms and chronicling of knowledge relating to biological diversity.

Explanation- For the purposes of this sub-section–

(a) "cultivar" means a variety of plant that has originated and persisted under cultivation or was specifically bred for the purpose of cultivation;

(b) "folk variety" means a cultivated variety of plant that was developed, grown and exchanged informally among farmers;

(c) "landrace" means primitive cultivar that was grown by ancient farmers and their successors.

(2) The National Biodiversity Authority and the State Biodiversity Boards shall consult the Biodiversity Management Committees while taking any decision relating to the use of biological resources and knowledge associated with such resources occurring within the territorial jurisdiction of the Biodiversity Management Committee.

(3) The Biodiversity Management Committees may levy charges by way of collection fees from any person for accessing or collecting any biological resource for commercial purposes from areas falling within its territorial jurisdiction.

Grants to Local Biodiversity Fund

42. The State Government may pay to the Local Biodiversity Funds by way of grants or loans such sums of money as the State Government may think fit for being utilized for the purposes of this Act.

Constitution of Local Biodiversity Fund

43.(1)There shall be constituted a Fund to be called the Local Biodiversity Fund at every area notified by the State Government where any institution of self-government is functioning and there shall be credited thereto-

(a) any grants and loans made under section 42;

(b) any grants or loans made by the National Biodiversity Authority;

(c) any grants or loans made by the State Biodiversity Boards;

(d) fees referred to in sub-section (3) of section 41 received by the Biodiversity Management Committees;

(e) all sums received by the Local Biodiversity Fund from such other sources as may be decided upon by the State Government.

Application of Local Biodiversity Fund

44.(1)Subject to the provisions of sub -section (2), the management and the custody of the Local Biodiversity Fund and the purposes for which such Fund shall be applied, be in the manner as may be prescribed by the State Government.

(2) The Fund shall be used for conservation and promotion of biodiversity in the areas falling within the jurisdiction of the concerned local body and for the benefit of the community in so far such use is consistent with conservation of biodiversity.

Annual Report of Biodiversity Management Committees

45. The person holding the custody of the Local Biodiversity Fund shall prepare, its annual report, giving a full account of its activities during the previous financial year, and submit a copy thereof to the concerned local body.

Audit of Accounts of Biodiversity Management Committees

46. The accounts of the Local Biodiversity Fund shall be maintained and audited in such manner as may, in consultation with the Accountant-General of the State, be prescribed and the person holding the custody of the Local Biodiversity Fund shall furnish, to the concerned local body, before such date as may be prescribed, its audited copy of accounts together with auditor's report thereon.

Annual Report, etc, of the Biodiversity Management Committee to Be Submitted to District Magistrate

47. Every local body constituting a Biodiversity Management Committee under sub-section (1) of section 41, shall cause, the annual report and audited copy of accounts together with auditor's report thereon referred to in sections 45 and 46, respectively and relating to such Committee to be submitted to the District Magistrate having jurisdiction over the area of the local body.

Chapter-XII Miscellaneous

This covers clauses 48 to 65 and these are mostly related to administrative issues, powers of central government, powers of state governments, court proceeding and penalties, and are not included here. However, for details, if required, NBA (2004) may be referred.

8

DUS Testing and Registration of Crop Varieties

In order to provide for the establishment of an effective system for protection of plant varieties, plant breeders' rights and the farmers' rights and to encourage the development of new varieties of plants it has been considered necessary to put in place a legal protection mechanism the Protection of Plant Varieties and Farmers' Rights Act, 2001 has been enacted in India (www.plantauthority.gov.in). For the purposes of this Act, Protection of Plant Varieties and Farmers' Rights Authority has been established and is located at NASC Complex, DPS Marg, Opp - Todapur, New Delhi-110 012.

Plant Variety Protection Registration Process

The details are available at the website (www.plantauthority.gov.in). The varietal registration process is as follows.

Reception

Reception desk of the authority receives application and allots PVP number and coupon to the applicant. Only completed application with all enclosures, registration fee (non-refundable) with requisite quantity of seed in officially sealed packing is accepted. Vegetatively propagated planting material is submitted directly to the respective crop DUS (distinctness, uniformity, stability) centre as per details available on website within 10 days of submitting the application with exception in case of perennial crops as listed on website where DUS testing will be done at site and no propagating material will be deposited either at the authority counter or at the designated DUS testing centre.

PVP Application Number

This number allotted by the authority shall facilitate updating of record, and printing of acknowledgement.

Acknowledgement

For application received by hand at headquarter/branch office before noon, acknowledgment can be collected between 5.00 to 5.30 pm in exchange of coupon. For applications received in the afternoon, acknowledgement can be collected on the next working day between 10.00 and 10.30 am. For applications received through post/courier, acknowledgement shall be posted latest by next working day by dispatch section.

Transferring the Application to Concerned Registrar and Examination of Application

Queries raised by Registrar to be issued within 7 days to the applicant for compliance within 15 days. Within next 7 days, the Registrar shall examine the reply and if acceptable, will ask for depositing DUS testing fee. If fee not received within 15 days, application with seed shall be returned to the applicant.

Allotment of Registration (REG) Number

If the application is accepted, the REG number shall be informed to the applicant and published in Plant Variety Journal (PVJ). REG number shall be the reference number of the proposed variety in all correspondence.

Dispatch of Seed to the DUS Centres

Registrar shall dispatch the seed to the concerned DUS testing centres as decided by the Plant Authority at least 15 days before the sowing date. In case of vegetatively propagated planting material, specific instruction shall be sent to the concerned DUS testing centres based on database. Crop-wise, DUS testing centres are available on the website of Plant Authority. DUS testing centres for major field crops are listed as follows.

Cereals

Crop	DUS Test Centres				
.	I	II	III	IV	V
Rice	IIRR, Hyderabad	NRRI, Cuttack	IARI, Karnal	AAU, Jorhat	TNAU, Coimbatore
Wheat	DWR, Karnal	IARI, Indore	CSUAT, Kanpur	UAS, Dharwad	PAU, Ludhiana
Maize	IMR, PAU Campus, Ludhiana	ANGRAU, Hyderabad			
Sorghum	IIMR, Hyderabad	MPKV, Rahuri	GBPUA&T, Pantnagar		
Pearl millet	MPKV, Rahuri	Project Coordinating Unit of AICRP (Pearl Millet), Jodhpur			

Pulses

Crop	DUS Test Centres			
.	I	II	III	IV
Chickpea	IIPR, Kanpur	CCS, HAU, Hisar	PDKV, Akola	
Mungbean	MULLaRP, IIPR, Kanpur	ANGRAU, Hyderabad	.	
Urdbean	MULLaRP, IIPR, Kanpur	ANGRAU, Hyderabad		
Field pea	MULLaRP, IIPR, Kanpur	JNKVV, Jabalpur		
Rajmash	MULLaRP, IIPR, Kanpur	IIVR, Varanasi	VPKAS, Almora	
Lentil	MULLaRP, IIPR, Kanpur	JNKVV, Jabalpur	.	
Pigeon Pea	MULLaRP, IIPR, Kanpur	PDKV, Akola	.	

Oilseeds

Crop	DUS Test Centres			
	I	II	III	IV
Mustard	DRMR, Bharatpur	CSAUA&T, Kanpur		
Groundnut	DGR, Junagarh	TNAU, Coimbatore		
Soybean	VPKAS, Almora	IISR, Indore	UAS, Dharwad	
Sunflower	DOR, Hyderabad	TNAU, Coimbatore		
Safflower	DOR, Hyderabad	PDKV, Akola		
Castor	DOR, Hyderabad	JAU, Jam Nagar		
Sesame	PC, Sesame and Niger, Jabalpur			
Linseed	PC,CSAU&T, Kanpur	JNKVV, Jabalpur		

Fibers

Crop	DUS Test Centres				
	I	II	III	IV	V
Cotton	CICR, Nagpur	UAS, Dharwad	HAU, Hisar	CICR, Coimbatore	PAU, Ludhiana
Jute	CRIJAF, Barrack pore	CSRS, Bud			

DUS Testing

DUS centres will test the candidate variety as per DUS guidelines prescribed for the crop by Plant Authority. If, at the end of the first season, it is noted that an essential trait is expressing significant distinctiveness between the locations, then Registrar/his nominee along with a representative of the applicant has to visit the sites for consensus decisions during the second season of the testing for DUS characters.

DUS Testing for Essentially Derived Varieties (EDV)

For testing of EDV hybrids or varieties, DUS characterization is done along with original hybrid/variety (initial hybrid/variety) and in case of hybrid EDV, essentially derived parents and original parents for one year at two locations.

Data Analysis

After DUS testing tabulated and certified pooled data from centres is submitted to the Plant Authority Registrar by the Principal Investigator (PI) of the DUS testing within four months of harvest. Registrar takes final decision within next 15 working days on the candidate variety.

Pre-Grant Opposition

The decision of Registrar along with all the details is published in Plant Variety Journal (PVJ) of the Plant Authority for inviting pre-grant opposition. If no pre-grant opposition is received within the stipulated time as per Section 21 (2), check-list will be prepared by the Registry within two weeks and Registration Certificate will be issued to the applicant.

Certificate of Registration

A certificate of registration is issued by Plant Authority in favour of the registered variety with following details.

Certificate of Registration No.......

(1) Registration Number and date of grant.

(2) Name and address of applicant or breeder in whose name the certificate has been issued or registered.

(3) Denomination of the variety.

(4) Name of:

Family

Genus

Species

Variety and common name

(5) Parentage and geographical location of the variety.

(6) Details of the distinguishing features or the characteristics.

(7) In case of "essentially derived variety□, the details of the essential variety from which "essentially derived variety" is claimed to have been derived.

(8) Name and address of the contributor, nature and amount of the contribution or the community knowledge used in the development of the plant variety.

(9) Terms and conditions of the agreement, if any, entered into between the breeder and contributor.

(10) If the variety is sold or otherwise disposed of, details thereof.

Plant Variety Registration Application Form

Common application form applicable to all the crops is available on Plant Authority website. Lot of information is to be filled as per DUS testing guidelines applicable to specific crops and these guidelines are also available on the Plant Authority website. For illustration, the application and DUS guidelines as applicable to maize are reproduced as follows (www.plantauthority.gov.in).

I. Subject

These test guidelines shall apply to all varieties, hybrids and parental lines of maize *(Zea mays* L.).

II. Seed Material Required

1. The Protection of Plant Varieties and Farmers' Rights Authority (PPV & FRA) shall decide when, where and in what quantity and quality of the seed material are required for testing a variety denomination applied for registration under the Protection of Plant Variety and Farmers' Rights (PPV & FR) Act, 2001. Applicants submitting such seed material from a country other than India shall make sure that all customs and quarantine requirements stipulated under relevant national legislations and regulations are complied with. The minimum quantity of the seed to be provided by the applicant shall be 3000 gram in the case of the candidate variety or hybrid and 1500 gram for each of the parental lines of the hybrid. Each of these seed lots shall be packed and sealed in ten equal weighing packets and submitted in one lot.
2. The seeds submitted shall have the following standards for germination, moisture content and physical purity.
   a. **Germination capacity:** i. Inbred lines and single cross hybrids: 80% (minimum) ii. Varieties and double cross hybrids: 90% (minimum)
   b. **Moisture content:** 8-10 % (maximum)
   c. **Physical purity**: 98% (minimum)
3. The applicant shall also submit along with the seed a certified data on germination test made not more than one month prior to the date of submission. It also shall possess the highest genetic purity, uniformity, sanitary and phytosanitary standards.
4. The plant material shall not have been subjected to any chemical and bio-physical treatment.

III. Conduct of Tests

1. The minimum duration of the DUS tests shall normally be at least two independent similar growing seasons.
2. The test shall normally be conducted at least at two test locations. If any essential characteristics of the candidate variety are not expressed for visual observation at these locations, the variety shall be considered for further examination at another appropriate test site or under special test protocol on expressed request of the applicant.
3. The field tests shall be carried out under conditions favouring normal growth and expression of all test characteristics. Each test shall include

about 250 plants in the plot size and planting space specified below across three replications. Separate plots for observation and measurement can only be used if they have been subjected to similar environmental conditions. All the replications shall be sharing similar environmental conditions of the test location.

IV. Test Plot Design

Number of rows: i. Inbred lines and single cross hybrids: 4, ii. Varieties and other hybrids: 8, Row length: 6 m, Row to row distance: 75 cm, Plant to plant distance: 20 cm, Number of replications: 3

5. Observations shall not be recorded on plants in border rows.
6. Additional tests for special purpose shall be established by the PPV & FR Authority.

V. Methods and Observations

1. The characteristics described in the Table of characteristics (see Section VII) shall be used for the testing of varieties, inbred lines and hybrids for their DUS. 2. For the assessment of Distinctiveness and Stability, observations shall be made on (excluding out-crossed plants in inbred lines and plants obviously resulting from the selfing of a parental line in single cross hybrids) at least 30 plants for inbreds/single cross hybrids and 60 plants for varieties and other hybrids. 3. For the assessment of Uniformity of inbred lines and single-cross hybrids a population standard of 1% with an acceptance probability of 95% shall be applied. In the case of a sample of 100 plants, maximum number of variants allowed would be 3 in case of inbreds and single cross hybrids and 6 in case of other varieties and hybrids. For three-way cross hybrids, double-cross hybrids and open-pollinated varieties, the variability within the variety shall not exceed the variability of comparable varieties already known. 4. All observation on ear shall be made on the upper well-developed ear. 5. For the assessment of all colour characteristics, the latest Royal Horticultural Society (RHS) colour chart shall be used.

VI. Grouping of Varieties

1. The candidate varieties for DUS testing shall be divided into groups to facilitate the assessment of Distinctiveness. Characteristics, which are known from experience not to vary, or to vary only slightly within a variety and which in their various states are fairly evenly distributed across all varieties in the collection are suitable for grouping purpose.

2. The following characteristics are proposed to be used for grouping maize varieties:

a) Tassel: Time of anthesis (Characteristic 4)

b) Ear: Time of silk emergence (50% plants) (Characteristic 11)

c) Ear: Anthocyanin colouration of silks (Characteristic 12)

d) Plant: Length (Characteristic 15.1 and 15.2)

e) Ear: Type of grain (Characteristic 22)

VII. Characteristics and Symbols

1. To assess Distinctiveness, Uniformity and Stability, the characteristics and their states as given in the Table of characteristics (Section VII) shall be used.
2. Notes (1 to 9) shall be given for each state of expression for different characteristics for the purpose of electronic data processing.
3. Legend:

(*) Characteristics that shall be observed during every growing season on all varieties and shall always be included in the description of the variety, except when the state of expression of any of these characters is rendered impossible by a preceding phenological characteristic or by the environmental conditions of the testing region. Under such exceptional situation, adequate explanation shall be provided.

(+) See Explanation on the Table of characteristics in Section VIII. It is to be noted that for certain characteristics the plant parts on which observations to be taken are given in the explanation or figure(s) for clarity and not the colour variation.

(S) Characteristics may segregate in three-way cross hybrids and double cross hybrids with the effect that several states of expression occur side by side in a hybrid variety.

4. A decimal code number in the sixth column of Table of characteristics indicates the optimum stage for the observation of each characteristic during the growth and development of plant. The relevant growth stages corresponding to these decimal code numbers are described below.

Decimal Code for the Growth Stage

Stage Code	General Description
00	Dry seed
12	2 leaves unfolded
14	4 leaves unfolded
51	Inflorescence just visible
61	Beginning of anthesis
65	Anthesis halfway
71	Caryopsis watery ripe
75	Medium milk
85	Soft dough
92	Caryopsis hard (can no longer be dented by thumbnail)
93	Caryopsis loosening daytime

5. Type of assessment of characteristics indicated in column 7 of Table of characteristics is as follows.

MG: Measurement by a single observation of a group of plants or parts of plants

MS: Measurement of a number of individual plants or parts of plants

VG: Visual assessment by a single observation of a group of plants or parts of plants

VS: Visual assessment by observation of individual plant or parts of plant

III. Table of Characteristics

S.N.	Characteristics	States	Note	Example (variety/ line)	Stage of observations	Type of assessment
1	2	3	4	5	6	7
1 (+)	Leaf: Angle between blade and stem (just on leaf above upper ear)	Small (<45°) Wide (>45°)	3 7	HHM-2 Vivek-5	61	VG
2 (+)	Leaf: Attitude of blade (on leaf just above upper ear)	Straight Drooping	1 9	HHM-2 HKI-323	61	VG
3 (S)	Stem: Anthocyanin colouration of brace roots	Absent Present	1 9	HKI-163 HKI-327T	65-75	VS
4 (*)	Tassel: Time of anthesis (on middle third of main axis, 50% of plants)	Very early (<45 days) Early (45-50 days) Medium (50-55 days) Late (>55 days)	1 3 5 7	HKI-335 HKI-1025 HKI-323 HKI-1126	65	VG
5 (+) (S)	Tassel: Anthocyanin colouration at base of glume (in middle third of main axis)	Absent Present	1 9	HKI-1344 HKI-161	65	VS
6 (S)	Tassel: Anthocyanin colouration of glumes excluding base (in middle third of main axis)	Absent Present	1 9	HKI-209 HKI-161	65	VS
7 (S)	Tassel: Anthocyanin colouration of anthers (in middle third of main axis on fresh anthers)	Absent Present	1 9	HKI-209 HKI-161	65	VG
8	Tassel: Density of spikelets (in middle third of main axis)	Sparse Dense	3 7	HKI-1126 HKI-288-2	65	VG
9 (*) (+)	Tassel: Angle between main axis and lateral branches (in lower third of tassel)	Narrow (<45°) Wide (>45°)	3 7	CM-145	65	VG
10 (*) (+)	Tassel: Attitude of lateral branches (in lower third of tassel)	Straight Curved Strongly curved	1 5 9	HKI-193-1 HKI-323, 46 HKI-163	65	VG

S.N.	Characteristics	States	Note	Example (variety/ line)	Stage of observations	Type of assessment
11	Ear: Time of silk emergence (50% plants)	Very early)less than 48 days)	1	HKI-335	65	VG
		Early (48-53 days)	3	HKI-1025		
		Medium (53-58 days)	5	HKI-323		
		Late (>58 days)	7	HKI-1126		
12 (+)	Ear: Anthocyanin colouration of silk (on day of emergence)	Absent	1	HKI-1025	65	VG
		Present	9	HKI-323		
13	Leaf: anthocyanin colouration of sheath (below the ear)	Absent	1	HKI-163	71	VS
		Present	9	CM-300		
14	Tassel: Length of main axis above lowest side branch	Short (<20 cm)	3	HKI-1128	71	MS
		Medium (20-30 cm)	5	HKI-327T		
		Long (>30 cm)	7	HKI-1105		
15.1 (*)	Inbred lines only: Plant: Length (up to flag leaf0	Short (<120 cm)	3	HKI-1348-6-2	75	MS
		Medium (120-150 cm)	5	HKI-323		
		Long (>150 cm)	7	HKI-1128		
15.2 (*)	Hybrids and open-pollinated varieties only: Plant: Length (up to flag leaf only)	Short (<150 cm)	3	HM-1	75	MS
		Medium (150-210 cm)	5	HM-4		
		Long (181-210 cm)	7	HQPM-1		
		Very long (>210 cm)	9	African Tall		
16	Plant: Ear placement	Low	3	HKI-1011	75	MS
		Medium	5	HM-4		
		High	7	HQPM-1		
17	Leaf: Width of blade (leaf of upper ear)	Narrow (<8 cm)	3	HKI-323	75	MS
		Medium (8-9 cm)	5	HKI-295		
		Broad (>9 cm)	7	HKI-1126		
18 (*)	Ear: Length without husk	Short (<10 cm)	3	HKI-536	92	MS
		Medium (10-15 cm)	5	HKI-163		
		Long (>15 cm)	7	HQPM-1		

S.N.	Characteristics	States	Note	Example (variety/ line)	Stage of observations	Type of assessment
19	Ear: Diameter without husk (in middle)	Short (<4 cm)	3	HKI-323	92	MS
		Medium (4-5 cm)	5	HKI-327		
		Large (>5 cm)	7	THQPM-1		
20 (+)	Ear shape	Conical	1	HKI-1344	92	VG
		Conico-cylindrical	2	HKI-295		
		Cylindrical	3	HKI-1105		
21	Ear: Number of rows of grains	Few 8	3	HKI-163	92	MS
		Medium 10-12	5	HM-5		
		Many >14	7			
22 (*)	Ear: Type of grain (in middle third of ear)	Flint	1	HKI-1105	92	VG
		Semi-flint/semi-dent	2	HKI-1344		
		Dent	3	HM-5		
23 (*)	Ear: Colour of top of grain	White	1	CM-300	92	VG
		White with cap	2	HKI-1344		
		Yellow	3	HKI-1025		
		Yellow with cap	4	HKI-209		
		Orange	5	HKI-323		
		Red	6			
		Other (specify)	7			
24 (*)	Ear: Anthocyanin colouration of glumes of cob	White	1	HKI-163	93	VG
		Light purple	2	HKI-295		
		Dark purple	3	HKI-161		
25 (+)	Kernel: Row arrangement (middle of ear)	Straight	1	HM-2	93	VG
		Spiral	2	HM-5		
		Irregular	3	HKI-1344		
26	Kernel: Poppiness	Absent	1	HM-1	93	VG
		Present	9	Amber popcorn		

S.N.	Characteristics	States	Note	Example (variety/ line)	Stage of observations	Type of assessment
27	Kernel: Sweetness	Absent	1	HM-1	93	VG
		Present	9	Madhuri		
28	Kernel: Waxiness	Absent	1	HM-1	93	VG
		Present	9			
29	Kernel: Opaqueness	Absent	1	HM-1	93	VG
		Present	9	CML-142		
30 (+)	Kernel: Shape	Shrunken	1	Madhuri	93	VG
		Round	2	HKI-1342		
		Indented	3	HM-5		
		Toothed	4	HKI-1348		
		Pointed	5			
31	Kernel: 1000 kernel weight	Very small (<100 g)	1	Madhuri	93	MG
		Small (100-200 g)	3	HKI-1025		
		Medium (200-300 g)	5	HQPM-1		
		Large (>300 g)	7			

Note: HKI = HAU/Haryana Karnal Inbred, HM = Hybrid Maize, HQPM = Hybrid Quality Protein Maize

Application Pro-forma for Registration of Cultivars with Plant Authority

The applicable format is as follows.

APPLICATION FOR REGISTRATION OF NEW VARIETY UNDER PPV&FRA

REGISTRATION OF NEW SUNFLOWER VARIETY / PARENTAL

SUBMITTED BY

SUMBITTED ON ______________________________

"FORM 1

[See regulation 10]

APPLICATION FOR REGISTRATION OF NEW VARIETY UNDER PROTECTION OF PLANT VARIETY AND FARMERS' RIGHTS ACT, 2001

[*See section 18, other than essentially derived variety.*]

(Instruction to applicant: Wherever a box item appears against queries, please tick the relevant box and provide legibly written/typed response in other queries.)

1. Identity of the Applicant(s)

☐ INDIVIDUAL BREEDER

☐ SUCCESSOR OF BREEDER

☐ INSTITUTIONAL APPLICANT

☐ ASSIGNEE OF ANY OF ABOVE[1]

☐ CONVENTION COUNTRY[2]

☐ ANY OTHER[3]

1. *An assignee or legal representative applicant shall submit required proof of the right of making the application in accordance with rule 27.*
2. *The Gazette Notification of Government of India should cover the country of the applicant and that India enjoys such mutual privileges (see rule 31 of the Act)*

3. *For material developed by participatory plant breeding, attach documents to explicitly show the deal struck between the farmer and the plant breeder.*

2. Name(s) and Nationality of Applicant(s)

(a.) (If natural person): [Insert additional rows, if required]

1. Serial Number. -------------------------
2. Name -------------------------
3. Complete Address -------------------------
4. Nationality -------------------------

(b.) (If a legal person; for example a firm or company or institution)

Name:

Address of its seat or establishment:

(Registered office):

Year of Incorporation:

State whether the applicant legal person has non-Indian participation in capital or management:

☐ YES ☐ NO

If yes, identify the nationality:

(c.) Indicate the name and address of the natural person, being an employee of the legal person, who is duly authorized to represent the legal person (*example a director of a company or a partner of a firm*):

Name : ----------------

Designation :---------------

Address : ---------------

Telephone : --------

Fax :

E-mail :

3. Name and Address of the Person to whom Correspondence related to this application is to be sent: (Attached authorization in Form-PV-1)

Name : ----------------

Address :----------------

Pin : ----------------

Telephone : ----------

Fax : -------------------

E-mail : -------------------

4. General Information of the Candidate Variety:

   Common name of the Crop:

   Botanical name[4] :

   Family :

   Denomination (in block letters):

   [4]Botanical names mean the scientific name approved by the International Code for Nomenclature of Cultivated Plants, 2004.

5. Type of Variety (see chapter III of the Protection of Plant Varieties and Farmers' Rights Authority, 2003):

   ☐ NEW VARIETY

   ☐ EXTANT VARIETY

6. (a.) Classification of the Candidate Variety:

   ☐ TYPICAL VARIETY[5]

   ☐ HYBRID VARIETY*

   ☐ TRANSGENIC

   ☐ OTHER (SPECIFY)

   Typical variety means a variety, which is not a hybrid or an essentially derived variety and normally propagated by using propagules saved from previous crop production cycles (Example: pure lines including parental lines/composite varieties or vegetative propagated varieties).

   *The hybrid and not a transgenic hybrid. In case of transgenic attach copy of the Genetic Engineering Approval Committee clearances for cultivation and seed productions (b.) What is (are) the Distinctness, Uniformity, Stability features on the basis of which registration is sought. Explain in detail the group characters (see specific guidelines for details). Attach 'Technical Questionnaire' sheet with all needed details duly signed with seal.

'Technical Questionnaire'---------------

(c.) If new variety is a transgenic attach clearance on Bio-safety from Ministry of Environment and Forests.

7. Names and Addresses of Breeder(s) who has/have bred the Candidate Variety*:

Name : ---------------------

Address : ---------------------

Telephone : ---------------------

Fax : ---------------------

E-mail : ---------------------

Nationality : ---------------------

**In case of more than one breeder, mention all names as (ii), (iii) and so on in the above format. If required insert extra page.*

(a.) Details of all other earlier applications made on the candidate variety in convention countries or other countries (if applicable):

Variety denomination: ------------------

Nature of right applied for: Plant breeder' rights ☐ Patent ☐

Filing Date

(Attach evidence): ____________________

Name of Country: ____________________

Name of Authority: ____________________

Application Number: ____________________

Status of Application: ☐ Under process ☐ Approved ☐ Rejected

(*If required, repeat the above for each applicable country and attach separate sheet*)

(b.) Priority is now claimed in respect of the earliest application for a candidate variety of said denomination (if applicable): -----------------

In (country): ---------------------------

On (date of application): ---------------------------

8. Has the candidate variety been commercialized or otherwise exploited?

☐ YES ☐ NO

If yes, please indicate the following:

Date of the first sale of the variety :______________________________

Country(ies) where Protection is made:______________________________

Denomination used : ______________________________

Trademark used, if any : ______________________________

Variation in important trait with respect to first filing:---------------

9. (a) If the candidate variety is a hybrid, state whether all the parental lines required for the repeated propagation of the hybrid are bred exclusively by the applicant(s):

☐ YES ☐ NO[6]

[6]If no, mention which of the parental line is outsourced, whether letter of agreement is obtained for each of the outsourced protected parental lines in compliance with Section 30 of the Protection of the Plant Varieties and Farmers' Rights Act and also provide following information on each of them:

Parental line (S): -----------------

Denominations 7:----------------

Source: ---------------------------

Authorization letter obtained: ☐ Attached ☐ Not attached

[7]Denomination should not be altered from what was used at the source. Information on source may include name of breeder or institution or farmer or farming community who had bred and maintained the parental line. Repeat above information for additional applicable parental line.

(b) State if any Farmers' Variety or Variety of Common Knowledge or variety in public domain is used as parental line for the repeated propagation of the hybrid:

☐ YES ☐ NO

If yes, give following details:

Denomination:---------------

Geographical Source:--------

Details of Attribution (Origin):------------

Details of Owner Farmer /Village Community/ Institution/ Organization: -------------

(c) The Protection of Plant Varieties and Farmers' Rights Act, 2001 provides access to benefit sharing to farmers who have conserved the genetic resource that has contributed towards variety development. In this particular case what sort of farmer/community recognition the Applicant has planned?

10. In case exotic germplasm was used in the derivation of the variety or hybrid, give details: ------------------------

11. Details on the payment of application fee and Distinctness Uniformity Stability testing fee:

Fees Type	Amount (Rs.)	In words	DD. No	Date	Bank & Branch
Application	------	----------	-------	------	--------
Registration	-------	----------	--------	-------	---------

(Signature of the Applicant)

Declarations

I hereby apply for the grant of registration of the candidate variety with the above said denomination and I am are conversant with the Protection of Plant Varieties and Farmers' Rights Act, 2001 and Rules thereof related to this application.

I hereby declare that no person other than the person or persons mentioned in this application has been involved in the breeding, or discovery or development of the candidate variety.

I hereby declare that the candidate variety complies with the sub- section (3) of section 29 of Protection of Plant Varieties and Farmers Rights Act, 2001.

I hereby declare that I/we have not applied for or received a trademark for the said denomination of the variety.

I hereby attach an affidavit in compliance with clause (C) of sub- section (1) of section 18 of Protection of Plant Varieties and Farmers' Rights Act, 2001.

I hereby declare that the information given in this application for the registration of the above said candidate variety, including annexure and all supporting documents are complete, true and correct to the best of my knowledge, information and belief and no information has been willfully concealed.

I hereby declare that genetic material or parental material acquired for breeding, evolving or developing the variety has been lawfully acquired.

I hereby declare that I shall abide by all the provisions and guidelines of Protection of Plant Varieties and Farmers' Rights Act, 2001.

Place:________________

Date: ________________

Signature of Applicant[8]

[8]Wherever the applicants are more than one person each applicant has to sign. In the case of authorised application or application by assignees, such person(s) authorised or assigned shall sign

Following are the attachments (duly signed/seal) submitted along with the application: (note that wherever signature is affixed in the application or attachments, all such signatures shall be in the original):

(a) Complete application: ----------------

(b) Document of authorization in Form PV-1:-------------

(c) Document of assignment in Form PV-2: --------------

(d) Documents in support of (b) and (c) as given above: -------------

(f) Affidavit that the Terminator Technology and the Genetic Use Restriction Technology is not involved:---------------

(g) Copy of document on filing date (vide column 8A):----------

(h) Copy of letter of agreement (vide column 10A):-------------

(i) Technical Questionnaire for the Candidate variety:------------

(j) If the applicants by virtue of succession or an assignment of the right to apply for registration attach a proof to show the right to application as stipulated in sub -section (3) of section 18 of the Protection of Plant Varieties and Farmers' Rights Act, 2001:-----------

(k) In case of Convention Country applicant attach complete details on the variation in the important trait with respect to first filing as enclosure:------------

(l) In case of Convention Country applicant provide information whether the variety has been sold or otherwise disposed of within or outside the convention Country with details thereof:--------------

(m) In case of transgenic relevant Genetic Engineering Approval Committee clearances and approvals:--------------

(n) Fees as applicable (Enclosed two DDs for Rs.-------/- & Rs.----------)

If felt necessary, attach colour pictures of specific characteristics used for establishing distinctness. Please sign each page of the application and other document on the left margin.

Technical Questionnaire

This questionnaire has been prepared for a hypothetical variety of sunflower so that the readers are acquainted with one more crop regarding filing of application for PPVFRA registration and the required information to be submitted.

1. Name of the company:--------------
2. Year of Establishment: ------------
3. If registered company under Company's Act 1956 (Give details): -
4. Location of corporate office and address: ----------------------
5. Tel/fax/e-mail: --------------
6. Name of candidate variety: Sunflower variety V-1.

   a) Has it been released in any Convention Country earlier?

      ☐ Yes ☐ No

   If yes give complete details in column number 13

   b) Pedigree/genealogy: ---------------

   c) Breeding of candidate variety: ----------------

(i) Origination: -------------

(ii) Controlled pollination/open pollination/induced mutation/spontaneous mutation/introduction/selection/seedling selection/any other ,specify :

Name of the parental material, characteristics of the parental material, distinguishable from the candidate variety should be given in a table. Just for illustration, hypothetical information is given in Table 8.1. This table illustrates what is the required information and how they are filled-up. The characters and grouping have been taken from sunflower-DUS-guidelines as available on Plant Authority website.

Table 8.1: Distinguishing characteristics of candidate variety V-1 of sunflower (hypothetical variety)

S.N.	Group of characteristics	Variety V-1
1	Plant: Time of flowering (DUS #2)	Early (3)
2	Leaf: Serration of leaf (DUS #7)	Medium (5)
3	Plant: Height (DUS #24)	Medium tall (5)
4	Plant: Branching (DUS #25)	Present (9)
5	Seed coat: Stripes (DUS # 31)	Present (9)

Sunflower variety-1: Plant: (hypocotyl anthocyanin pigmentation-absent to very weak), time of flowering: (early), leaf: (size-small, shape-triangular, colour-green, blistering-medium, serration-medium), stem pigmentation; -absent, disk floret: (colour-yellow), anthocyanin pigmentation of stigma: (absent), ray floret number: (few), head: (head attitude- vertical, shape on grain side-flat, diameter-small, natural position of closest lateral head to the central head- below), branching: (present),), plant height: (medium tall), seed: (length-short, shape-ovoid elongated, base colour-black, stripes –present and black in colour).

(iii) Breeding technique/procedure used

Pedigree method of breeding

(iv) Selection criteria used

Economically important traits

(v) Stage of selection and multiplication

The selection of the candidate parental line variety was done after testing its comparative agronomic performance in breeding block at the research station RS-1. The candidate variety was found to be better than the existing parental lines of commercial marketed hybrids and national checks and showed tolerance to major diseases like downy mildew, rust, bud necrosis virus and Alternaria blight of sunflower. The candidate line variety has been multiplied following standard seed multiplication stages such as nucleus seeds/ breeder seeds/foundation seeds and can be used as parental cultivar in future sunflower hybrids.

(vi) Location where breeding was conducted: Research station, RS-1.

7. Particulars of comparative trial conducted by the applicant, if any

Information on the location, place, period and year/month of comparative trial conducted method of cultivation such as open field, facilities, planting, potting etc., scale of cultivation, reference varieties used, criteria for choice of the reference varieties, design of experiment, method of analysis of variance experimental error where applicable, and other details.

NOTE: Applicant may, furnish data, tables, copy (ies) of publication(s) related to the details of breeding, comparative trial and comparative data in addition to table of characteristics of candidate and reference varieties. This information provided under this item will not be published by the Authority but will be used to facilitate examination of candidate variety.

8. Characteristics of the candidate variety

a) (i) Give group characters

Table 8.2: Group characters of candidate sunflower variety (V-1)

Group characteristics		Remarks- measured values etc.
Plant	Time of flowering (DUS #2)	Early (3)
Leaf	serration (DUS # 7)	Medium (5)
Plant	height (DUS # 24)	Medium tall (5)
Plant	Branching (DUS # 25)	Present (9)
Seed	Stripes (DUS # 31)	Present (9)

(ii) Distinguishing characteristics (descriptive or elaborate): These are given in Table 8.3 based on sunflower DUS-guidelines (www. plantauthority.gov.in)

Table 8.3: Descriptive or elaborate characteristics of the candidate variety (V-1)

Characteristics	Characteristics value of candidate variety									Remarks- measured value etc. of candidate variety
	1	2	3	4	5	6	7	8	9	V-1
Hypocotyl: anthocyanin pigmentation	√									Absent(1) / very weak
Plant: Time of flowering			√							Early (3)
Leaf: Colour					√					Green (5)
Leaf: Blistering					√					Medium (5)
Leaf: Serration					√					Medium (5)
Leaf: Petiole anthocyanin pigmentation	√									Absent(1) / very weak
Disk flower: Anthocyanin pigmentation of stigma	√									Absent (1)
Plant: Natural position of closest lateral head to the central head (end of flowering)									√	Below (9)
Head: Attitude		√								Vertical (2)
Head: Diameter			√							Small (3)
Head: Shape on grain side		√								Flat (2)
Plant: Height			√							Medium tall (5)
Plant: Type of branching		√								Overall (2)
Seed coat: Base colour				√						Black (4)
Seed coat: Stripes									√	Present (9)

b) Table of characteristics between candidate denomination and reference variety

Please give replicated values for all of its distinguishing and other description for important characteristics along with the corresponding average values of the references varieties.

NOTE: Two or more reference varieties should be compared with the candidate variety in the characteristics table, including one deemed to be the most similar variety and other(s) as obvious/similar as possible. If you provide this information it will facilitate the Authority in their DUS test further in examination of the candidate variety.

Reference varieties: VR-2 and VR-3

Table 8.4: Distinguishing characters between candidate denomination (V-1) and reference varieties

S. No.	Characteristics	Characteristics value of candidate variety									Remarks/ measured value etc.	Characteristics value of reference variety	
		1	2	3	4	5	6	7	8	9	V-1	V-2	V-3
1	Hypocotyl: anthocyanin pigmentation	√									absent (1) to very weak	strong (9)	strong (9)
2	Plant: Time of flowering			√							early (3)	medium (5)	medium (5)
3	Leaf: Size (cm)	√									small (1)	small (1)	medium (3)
4	Leaf: Shape		√								triangular (2)	lanceolate(1)	cordate (3)
5	Leaf: Colour					√					green (5)	green (5)	light green (3)
6	Leaf: Blistering					√					medium (5)	absent (1)	strong (7)
7	Leaf: Serration					√					medium (5)	fine (3)	medium (5)
8	Leaf: Angle of lateral veins	√									acute (1)	acute (1)	obtuse (2)
9	Leaf: Orientation of blade		√								drooping (2)	drooping (2)	drooping (2)
10	Leaf: Petiole anthocyanin pigmentation	√									absent (1)	present (9)	present (9)
11	Stem: Pigmentation	√									absent (1)	present (9)	absent (1)
12	Ray floret: Number			√							few (3)	many (7)	medium (5)
13	Ray floret: Shape	√									elongated (1)	elongated (1))	elongated (1)
14	Ray floret: Colour		√								yellow (2)	yellow (2)	yellow (2)
15	Disk floret: Colour	√									yellow (1)	yellow (1)	yellow (1)
16	Disk flower: Stigma anthocyanin colour	√									absent (1)	medium (5)	medium (5)
17	Disk floret: Pollen colour		√								yellow (2)	yellow (2)	yellow (2)
18	Bract: Shape	√									elongated (1)	elongated (1)	elongated (1)
19	Bract: Anthocyanin pigmentation	√									absent (1)	absent (1)	absent (1)

S. No.	Characteristics	Characteristics value of candidate variety								Remarks/ measured value etc.	Characteristics value of reference variety	
20	Plant: Natural position of closest lateral head to the central head (end of flowering)-branched								√	below (9)	above (1)	below (9)
21	Head: Attitude	√								vertical (2)	turned down (4)	turned down (4)
22	Head: Diameter		√							small (3)	Small (3)	small (3)
23	Head: Shape on grain side		√							flat (2)	flat (2)	flat (2)
24	Plant: Height				√					Medium tall (5)	short (3)	Medium tall (5)
25	Plant: Branching								√	present (9)	present (9)	present (9)
26	Plant: Type of branching	√								overall (2)	overall (2)	overall (2)
27	Seed: Length (cm)		√							short (3)	medium (5)	short (3)
28	Seed: Shape	√								ovoid elongated (2)	elongated (1)	ovoid elongated (2)
29	Seed: Weight (100 seeds weight in g)		√							low (3)	low (3)	low (3)
30	Seed coat: Base colour			√						black (4)	black (4)	brown (3) with mottling
31	Seed coat: Stripes								√	present (9)	Absent (1)	present (9)
32	Seed coat: Colour of stripes	√								gray (2)	NA (non-stripped)	brown (3)
33	Seed: Hull content (%)		√							low (3)	medium (5)	Low (3)
34	Seed: Oil content (%)						√			High (7)	medium (5)	medium (5)

9. Characteristics of the reference varieties

a) Most similar variety

(i) Denomination: V-2

(ii) Basis of choice of this variety for comparison: The reference variety is of same maturity as the candidate variety and agronomic requirements are similar. Further, the reference variety is released and in public domain.

(iii) Distinguishable characteristics

The reference variety V-2 has distinguishing characteristics with candidate variety in terms of - leaf colour-green, leaf angle of lateral veins-acute, branching-over all, ray floret shape-elongated, head size-small, head shape-flat, and seed coat base colour-black.

Table 8.5: Distinguishing characters between candidate denomination V-1 and reference variety

Characteristics	Characteristics value of candidate variety									Remarks/ Measured value etc. of candidate variety	Characteristics value of reference variety
	1	2	3	4	5	6	7	8	9	V-1	V-2
Hypocotyl: anthocyanin pigmentation	√									Absent(1) / very weak	Strong(9)
Plant: Time of flowering			√							Early (3)	Medium (5)
Leaf: Blistering					√					Medium (5)	Absent (1)
Leaf: Serration					√					Medium (5)	Fine (3)
Leaf: Petiole anthocyanin pigmentation	√									Absent(1) / very weak	Present(9)
Disk flower: Anthocyanin pigmentation of stigma	√									Absent (1)	Medium (5)
Plant: Natural position of closest lateral head to the central head (end of flowering)-Branched									√	Below (9)	Above (1)
Head: Attitude		√								Vertical (2)	Turned down (4)
Seed coat: Stripes									√	Present (9)	Absent (1)

b) Other reference variety

(i) Denomination: V-3

(ii) Basis of choice of this variety for comparison

The reference restorer lineV-3 is a male of hybrid H-1 which is already in public domain and has common agronomy with the candidate variety for same maturity group and common area of adaptation.

(iii) Distinguishable Characteristics

The reference variety-3 has distinguishing characteristics similar with candidate variety V-1 in terms of leaf size-serration-medium, disk floret colour-yellow, natural position of closest lateral head to the central head-below, head size-small, head shape-flat, branching-overall.

Table 8.6: Distinguishing characters between candidate denomination (V-1) and reference variety

Characteristics	Characteristics value of candidate variety									Remarks/ Measured value etc. of candidate variety	Characteristics value of reference variety
	1	2	3	4	5	6	7	8	9	V-1	V-3
Hypocotyl: Anthocyanin pigmentation	√									Absent (1) / very weak	Strong (9)
Plant: Time of flowering			√							Early (3)	Medium (5)
Leaf: Colour					√					Green (5)	Light green (3)
Leaf: Blistering					√					Medium (5)	Strong (7)
Leaf: Petiole anthocyanin pigmentation	√									Absent(1) / very weak	Present (9)
Disk flower: Anthocyanin pigmentation of stigma	√									Absent (1)	Medium (5)
Head: Attitude		√								Vertical (2)	Turned down (4)
Plant: Height			√							Short (3)	Medium (5)
Seed coat: Base colour				√						Black(4)	Brown (3)

10. Statement of distinctness of candidate variety: V-1

Table 8.7: Characteristics and values of candidate variety compared with the reference varieties

Characteristics	Characteristics value of candidate variety									Remarks/ measured value etc. of candidate variety	Characteristics value of reference variety	Characteristics value of reference variety
	1	2	3	4	5	6	7	8	9	V-1	V-2	V-3
Hypocotyl: Anthocyanin pigmentation	√									Absent(1) / very weak	Strong (9)	Strong (9)
Plant: Time of flowering			√							Early (3)	Medium (5)	Medium (5)
Leaf: Colour					√					Green (5)	Green (5)	Light green (3)
Leaf: Blistering					√					Medium (5)	Absent (1)	Strong (7)
Leaf: Serration					√					Medium (5)	Fine (3)	Medium (5)
Leaf: Petiole anthocyanin pigmentation	√									Absent (1) / very weak	Present (9)	Present (9)
Disk flower: Anthocyanin pigmentation of stigma	√									Absent (1)	Medium (5)	Medium (5)
Plant: Natural position of closest lateral head to the central head (end of flowering)-Branched									√	Below (9)	Above (1)	Below (9)
Head: Attitude		√								Vertical (2)	Turned down (4)	Turned down (4)
Head: Diameter			√							Small (3)	Small (3)	Small (3)
Head: Shape on grain side		√								Flat (2)	Flat (2)	Flat (2)
Plant: Height			√							Short (3)	Short (3)	Medium (5)
Plant: Branching		√								Overall (2)	Overall (2)	Overall (2)
Seed coat: Base colour				√						Black (4)	Black (4)	Brown (3)
Seed coat: Stripes									√	Present (9)	Absent (1)	Present (9)

Statement on Uniformity and Stability of Candidate Variety

Please give a brief statement describing any variation in the variety that may be regarded as part of its normal uniform or stable expression, which is predictable, capable of being described in clear terms and commercially acceptable. This should include description and frequency of any off-types, variants or mutations. In your opinion what should be the frequency of off types or any other describable variation beyond which the candidate variety shall be deem to be non-uniform. Also please point out which are the traits that may be particularly referred to as indicators to determine an unstable expression of the phenotype of candidate variety.

The characters such as leaf colour, plant height, head shape and size are influenced by g x e interactions. The expression of distinguishing traits can be found variable in some specific environmental conditions. Mostly soil fertility, moisture and environment influence maturity, head shape and size. The frequency of off – types can be in the range of 1-2 percent due to heritable / non-heritable variations.

12. Methods for maintaining the candidate variety

Please provide in a brief statement as to how the propagating material will be maintained throughout the duration of the plant breeder's right, and complete address where the variety will be maintained. This should include status of varieties that are not propagated by seeds including place and method of maintenance and storage of their vegetative material.

NOTE: *The holder of a plant breeder's right is responsible for ensuring that propagating material representative of the variety is maintained for the duration of the right.*

Basic seed like nucleus and breeder seed of this variety are maintained by the breeder under controlled conditions. The foundation seed is maintained and multiplied through sibbing under prescribed isolation for the crop species under supervision of a competent breeder.

13. Information on variety registered in Convention Countries: NA.

a. What were the grouping characters in that application for this candidate variety?- NA.

b. What was the Distinctness, Uniformity and Stability parameter on which it was registered? NA.

c. What is the variation in important trait with respect to first filing and the present one (Attach photograph)? NA.

d. Has the variety been withdrawn in the first filed country from cultivation or banned or from any of the subsequently released country? NA.

e. If so, the reasons (supplement with information)? NA.

I/We hereby declare that no person other than the person or persons mentioned in this application has been involved in the breeding, or discovery or development of this denomination.

Signature of Witness and Address Date..............................

The breeder/company should sign with date each page of the Technical Questionnaire.

9

All India Coordinated Research Projects

A significant development in the field of agricultural research in India is the formulation of All India Coordinated Research Projects (**AICRPs**), initially for the improvement of agricultural crops, but later extended to all aspects of crop husbandry.

Brief History

Indian Council of Agricultural Research (ICAR) established the first Coordinated Crop Improvement Project on Maize (CCIPM) in 1957. Before inception of the coordinated programme, state economic botanists used to undertake maize research in India as part of their crop improvement activity. Dr. E.W. Sprague, a decorated maize scientist from the Rockefeller Foundation (RF), was the first coordinator of the first of its kind centralized crop improvement programme in India. Dr. N.L. Dhawan, a distinguished plant breeder from ICAR superseded Dr. Sprague. The CCIPM was further strengthened in 1963 through PL 480 funds by appointing new research staff and it was renamed as All Indian Coordinated Maize Improvement Project (AICMIP). It was further upgraded to Directorate of Maize Research (DMR) in January 1994. On 9 February 2015, DMR was upgraded to ICAR-Indian Institute of Maize Research (ICAR-IIMR) to further consolidate the maize research programme in the country. The Headquarter of ICAR-IIMR was shifted from New Delhi to Ludhiana in 2016. The off-season nursery, Winter Nursery Centre (WNC) was established at Amberpet, Hyderabad in 1962 to accelerate the maize improvement programme of north Indian maize research centres where winter temperature is not conducive for maize cultivation. Subsequently, the WNC was shifted to Rajendranagar, Hyderbad in 2008. Realizing the growing importance of winter (*rabi*) maize in eastern India and to cater the need of seed production a Regional Maize Research and Seed Production Centre (RMRSPC) was established at Kushmahaut, Begusarai, Bihar in 1997.

All India Coordinated Research Project (Maize)

Since this was the first project, this is taken as a model project to outline the activities of any all India coordinated research project. The descriptions as given for this project under various heads will be applicable to other crops also except that the crop specific requirements will change but the broad outline will remain the same.

Overview

All India Coordinated Research Project (AICRP) on maize was started with the basic objective to develop and disseminate superior cultivars and production/protection technologies across maize growing regions of the country. Based on agro-climatic conditions, the country has been divided into five zones constituting of 32 centres. AICRP organizes multidisciplinary, inter-institutional systematic testing of newly developed cultivars (OPs and hybrids) from both private and public sectors. In addition, it also develops and validates production and/or protection technologies for different agro-climatic zones of the country.

Zone-wise Distribution of Centres in AICRP

The country is divided into various agro-climatic zones as per crop requirement where there is considerable homogeneity within the zone but there are climatological and topographical differences between the zones. The idea behind the creation of zones is to develop and standardize crop agro-technologies (varieties, production and crop protection technologies, seed production research, etc.) specifically suited to a particular zone. It is obvious that the same technology cannot be equally effective for the entire country and hence creation of zones based on climate and topography is an integral part of any AICRP. Within zones, centres (SAUs and ICAR crop based institutes) are selected and awarded funds from ICAR to carry out the experiments allotted to them under AICRP. For maize, the zones and the centres are as given in Table 9.1.

Table 9.1: Zones and centres under all India coordinated research project (maize)

Zone	Zone name	Centres and abbreviations
I	Northern Hill Zones (NHZ)	Srinagar (Sri), Kangra (Kan), Bajaura (Baj), Almora (Alm), Barapani (Bar), Imphal (Imp)
II	North West Plain Zone (NWPZ)	Ludhiana (Lud), Karnal (Kar), Pantnagar (Pan), Delhi (Del)
III	North East Plain Zone (NEPZ)	Baharaich (Bah), Varanasi (Var), Dholi (Dho), Sabour (Sab), Ranchi (Ran), Kalyani (Kal), Ambikapur (Amb), Bhubaneswar (Bhu), Gossaingaon (Gos)
IV	Peninsular Zone (PZ)	Rahuri (Rah), Kolhapur (Kol), Karimnagar (Kar), Hyderabad (Hyd), Peddapuram (Ped), Dharwada (Dha), Mandya (Man), Coimbatore (Coi), Vagarai (Vag)
V	Central Western Zone (CWZ)	Udaipur (Uda), Banswara (Ban), Godhara (God), Chhindwara (Chh)

Mandate of AICRP on Maize

- Develop superior hybrids and varieties combining high yield and acceptable quality of grain and fodder, wider adaptability and resistance to major pests, diseases and abiotic stress factors for each zone.
- Evolve appropriate crop management practices and formulate efficient maize-based cropping systems for sustainable maize production in each zone.
- Conduct investigations on key or potential pests and diseases of maize and identify and evolve elite sources of resistance to develop suitable integrated plant protection strategies for increasing the stability of production.
- Promote research and extension to meet local needs within each state through SAUs and other partners.

Objectives of AICRP on Maize

- Execution of strategic and applied research for genetic improvement in yield, quality and resistance to biotic and abiotic stresses.
- Development of varieties and hybrids with emphasis on single cross hybrids of different maturity durations to fit into *kharif*, *rabi* and spring cropping seasons and maize-based cropping systems for location-specific conditions in different parts of India.
- Development of efficient package of practices for increasing productivity.

- Tailoring maize for diversified uses for industry and special purposes in other sectors such as Quality Protein Maize, Baby Corn, Sweet Corn, Bio-Fuel, *etc.*
- Strategic research on post-harvest handling and value addition of maize.
- Maintenance, development and evaluation of germplasm for use in breeding.
- To organize breeder seed production programme for maize hybrids and cultivars.
- To organize on-farm research to reduce the yield gap between attainable and realized yield on farmers' fields.

List of All India Coordinated Research Projects (AICRP)

The list of ICAR-all India coordinated research projects on various crops numbering 60 and network projects (initial version of full-fledged AICRP) numbering 20 and few related smaller projects numbering 10 is given in Table 9.2 (icar.org.in).

Table 9.2: List of all India coordinated research projects, network projects and related projects under Indian Council of Agricultural Research

All India Coordinated Research Projects – 59	
1.	AICRP on Nematodes, New Delhi
2.	AICRP on Maize, New Delhi (shifted to Ludhiana)
3.	AICRP Rice, Hyderabad
4.	AICRP on Chickpea, IIPR, Kanpur
5.	AICRP on MULLARP, IIPR, Kanpur
6.	AICRP on Pigeon Pea, IIPR, Kanpur
7.	AICRP on Arid Legumes, Kanpur
8.	AICRP on Wheat & Barley Improvement Project, Karnal
9.	AICRP Sorghum, Hyderabad
10.	AICRP on Pearl Millets, Jodhpur
11.	AICRP on Small Millets, Bangalore
12.	AICRP on Sugarcane, Lucknow
13.	AICRP on Cotton, Coimbatore
14.	AICRP on Groundnut, Junagarh
15.	AICRP on Soybean, Indore
16.	AICRP on Rapeseed & Mustard, Bharatpur
17.	AICRP on Sunflower, Safflower, Castor, Hyderabad
18.	AICRP on Linseed, Kanpur
19.	AICRP on Sesame and Niger, Jabalpur
20.	AICRP on IPM and Biocontrol, Bangalore

21. AICRP on Honey Bee Research & Training, Hisar
22. AICRP –NSP (Crops), Mau
23. AICRP on Forage Crops, Jhansi
24. AICRP on Fruits, Bangalore
25. AICRP Arid Zone Fruits, Bikaner
26. AICRP Mushroom, Solan
27. AICRP Vegetables including NSP Vegetable, Varanasi
28 AICRP Potato, Shimla
29. AICRP Tuber Crops, Thiruvananthapuram
30. AICRP Palms, Kasaragod
31. AICRP Cashew, Puttur
32. AICRP Spices, Calicut
33. AICRP on Medicinal and Aromatic Plants including Betel Vine, Anand
34. AICRP on Floriculture, New Delhi
35. AICRP in Micro Secondary & Pollutant Elements in Soils and Plants, Bhopal
36. lAICRP on Soil Test with Crop Response, Bhopal
37. AICRP on Long Term Fertilizer Experiments, Bhopal
38. AICRP on Salt Affected Soils & Use of Saline Water in Agriculture, Karnal
39. AICRP on Water Management Research, Bhubaneswar
40. AICRP on Ground Water Utilisation, Bhubaneswar
41. AICRP Dryland Agriculture, Hyderabad
42. AICRP on Agro-meteorology, Hyderabad including Network on Impact Adaptation & Vulnerability of Indian Agri. to Climate Change
43. AICRP Integrated Farming System Research, Modipuram including Network Organic Farming
44. AICRP Weed Control, Jabalpur
45. AICRP on Agro-forestry, Jhansi
46. AICRP on Farm Implements & Machinery, Bhopal
47. All India Coordinated Research Project on Ergonomics and Safety in Agriculture, Bhopal
48. AICRP on Energy in Agriculture and Agro Based Industries, Bhopal
49. AICRP on Utilization of Animal Energy (UAE), Bhopal
50. AICRP on Plasticulture Engineering and Technologies, Ludhiana
51. AICRP on PHT, Ludhiana
52. AICRP on Goat Improvement, Mathura
53. AICRP- Improvement of Feed Sources & Nutrient Utilisation for raising animal production, Bangalore
54. AICRP on Cattle Research, Meerut
55. AICRP on Poultry, Hyderabad
56. AICRP-Pig, Izatnagar
57. AICRP Foot and Mouth Disease, Mukteshwar
58. AICRP ADMAS, Bangalore
59. AICRP on Home Science, Bhubaneswar

Network Projects – 20

1. All India Network Project on Pesticides Residues, New Delhi
2. All India Network Project on Underutilized Crops, New Delhi
3. All India Network Project on Tobacco, Rajahmundry
4. All India Network Project on Soil Arthropod Pests, Durgapura
5. Network on Agricultural Acarology, Bangalore
6. Network on Economic Ornithology, Hyderabad
7. All India Network Project on Rodent Control, Jodhpur
8. All India Network Project on Jute and Allied Fibres, Barrackpore
9. Network Project on Improvement of Onion & Garlic, Pune
10. Network Bio-fertilizers, Bhopal
11. Network Project on Harvest & Post Harvest and Value Addition to Natural Resins & Gums, Ranchi
12. Network Project on Animal Genetic Resources, Karnal
13. Network Project on R&D Support for Process Up-gradation of Indigenous Milk Products for Industrial Application, Karnal
14. Network Programme on Sheep Improvement, Avikanagar
15. Network Project on Buffaloes Improvement, Hisar
16. Network on Gastro Intestinal Parasitism, Izatnagar
17. Network on Haemorrhagic Septicemia, Izatnagar
18. Network Programme Blue Tongue Disease, Izatnagar
19. Network Project on Conservation of Lac Insect Genetic Resources, Ranchi
20. Network Project on Agricultural Bioinformatics and Computational Biology, New Delhi

Other Projects – 10

1. Technology Mission on Cotton (CICR, Nagpur)
2. Technology Mission on Jute (CRIJAF, Barrack pore)
3. Continuation, Strengthening and Establishment of Krishi Vigyan Kendras
4. Strengthening & Development of Higher Agricultural Education in India, New Delhi
5. Central Agricultural University, Imphal
6. Strengthening and Modernization of ICAR Headquarters
7. Intellectual Property Management & Transfer/Commercialisation of Agricultural Technology (Upscaling of Existing Component IPR HQ)
8. Indo-US Knowledge Initiative
9. National Agricultural Innovative Project, New Delhi
10. National Fund for Basic and Strategic Research, New Delhi

Operations of AICRPs for Crops

Considering the example of all India coordinated research project (maize) as model, it is made clear that all the ACRIPs meant for individual crops or even a group of crops operate on the same broad framework as outlined in case of maize. Every crop based AICRP has mission, vision and specific objectives centric to the crop for which the AICRP has been approved by the ICAR.

Broadly the crop based AICRPs have following objectives.

1. Plant genetic resources management including germplasm collection, characterization, documentation, evaluation, conservation and utilization by the centres under the AICRP and also to encourage germplam exchange among the centres.

2. Developing high yielding varieties/hybrids having resistance to major biotic stresses prevalent in the zone and also developing similar cultivars with tolerance to abiotic stresses.

3. Developing and standardizing crop production (nutrient management, water management, weed management) and crop protection (standardizing appropriate doses of fungicides, insecticides, etc.) technologies to boost the productivity of the cultivars developed by the breeders of the centre.

4. Carrying out seed production research as per need of the crop and solving emerging new problems, if any.

5. Dissemination of knowledge on appropriate cultivars and package of practices through KVKs, front line demos and the line department of agriculture of the concerned state governments.

The participating centres (main centres/sub-centres/voluntary centres) funded by the ICAR carry out the research activities/experiments as allotted to them by the AICRP. AICRP conducts an annual work shop/group meeting normally spread over three days at a particular state agricultural university/ICAR institute where representatives/designated scientists in the AICRP participate, discuss the results of the past one year, make specific recommendations and formulate the technical programme under each discipline for the next year. The proceedings are finalized and made available to the scientist in-charge of each AICRP centre and all other concerned as soft copy through mail.

One of the major recommendations pertains to identification of superior cultivars by a specific group constituted for this very purpose and this group examines the breeding trial data for past three years and also takes due note of agronomic, plant pathological and entomological data and identifies certain cultivars along with the zones for which the cultivars are recommended. Once these cultivars are identified by the crop work shop/group meeting, these become eligible to central varietal release for which the concerned crop breeder has to make release proposal in prescribed pro-forma and submit prescribed number of copies to the convener, central varietal release committee for release and notification of the cultivars.

There are parallel recommendations from agronomy, plant pathology and entomology disciplines regarding the appropriate crop production technology specific to the cultivars identified so that the famers get a complete technology package including cultivars and crop production and protection technology. Crop-wise breeder seed production responsibilities are also allotted to the participating centres during these workshops/group meetings.

The individual AICRPs are headed by a Crop Coordinator appointed by the ICAR. The Crop Coordinator is usually allotted office space in the premises of the National Research Centre/Directorate/Institute meant for the crop and operates independently from the Director of the NRC/Directorate/Institute of the crop. Over the years, considering emerging lack of coordination between the Crop Coordinator and the Director of the NRC/Directorate/Institute of the crop, now the new trend is emerging where the position of the Crop Coordinator is merged with that of the Director and then Director creates a Coordinating Cell within the institute, identifies a principal scientist level scientists along with some deporting staff and carries out the roles and responsibilities of the Crop Coordinator himself with the support of the Coordinating Cell. This arrangement is on increase. For details on individual crop based AICRP, www.icar.org.in may be referred.

In some crops, there is provision of Network Projects where a sub-group of centres is created to carry out the activities like full-fledged coordinated programmes till this Network Project becomes strong enough to be converted into a full-fledged AICRP. Typical example is All India Network Research Project on Onion and Garlic (AINRPOG) where before establishment of this Network project, onion and garlic used to part of all India coordinated research project (vegetable crops). There is also a trend to name AICRPs differently too for example all India coordinated research project (maize) is also referred to as all India coordinated maize improvement project and this is applicable to all the crops.

10

International Agriculture Research Centres on Field Crops

The 15 CGIAR (Consultative Group on International Agricultural Research) Research Centers are independent, non-profit research organizations, conducting innovative research. Home to more than 8,000 scientists, researchers, technicians, and staff, CGIAR research works to create a better future for the world's poor. Each Center has its own charter, board of trustees, director general, and staff. CGIAR Research Centers are responsible for hands-on research programmes and operations guided by policies and research directions set by the System Management Board.

International Agricultural Research Centres

Currently there are 15 centres as listed below (www.cgiar.org).

1. Africa Rice Centre
2. Centre for International Forestry Research (CIFOR)
3. International Centre for Agricultural Research in the Dry Areas (ICARDA)
4. International Crop Research Institute for the Semi-Arid Tropics (ICRISAT)
5. International Food Policy Research Institute (IFPRI)
6. International Centre of Tropical Agriculture (IITA)
7. International Livestock Research Institute (ILRI)
8. International Maize and Wheat Improvement Centre (CIMMYT)
9. International Potato Centre (CIP)
10. International Rice Research Institute (IRRI)
11. International Water Management Institute (IWMI)

12. Biodiversity
13. International Centre for Tropical Agriculture (CIAT)
14. World Agroforesty (ICRAF)
15. World Fish

Out of the above listed 15 centres, eight centres (African Rice Centre, ICARDA, ICRISAT, IITA, CIMMYT, CIP, IRRI and CIAT) are directly involved in improvement of major field crops and hence these eight centres are briefly described as follows.

1. Africa Rice Centre (AfricaRice)

AfricaRice is a CGIAR Research Center – part of a global research partnership for a food-secure future. It is also an intergovernmental association of African member countries. The Center was created in 1971 by 11 African countries. Today its membership comprises 27 countries, covering West, Central, East and North African regions, namely Benin, Burkina Faso, Cameroon, Central African Republic, Chad, Côte d'Ivoire, Democratic Republic of Congo, Egypt, Ethiopia, Gabon, the Gambia, Ghana, Guinea, Guinea Bissau, Liberia, Madagascar, Mali, Mauritania, Mozambique, Niger, Nigeria, Republic of Congo, Rwanda, Senegal, Sierra Leone, Togo and Uganda.

AfricaRice headquarters is based in Côte d'Ivoire. Staffs are located in Côte d'Ivoire and also in AfricaRice Research Stations in Benin, Liberia, Madagascar, Nigeria, and Senegal..

Major Research Programmes

Genetic Diversity and Improvement Programme

The turnover rate of rice varieties remains relatively low and older low-yielding varieties still dominate in large areas of Africa. Farmers are not benefiting quickly enough from genetic gains from recently improved varieties that are resilient to abiotic and biotic stresses, including those that can withstand heat, cold, drought, and flood that are forecast to become more frequent and intense with climate change.

The GDI programme is responsible for rice germplasm collection, storage, characterization, and improvement. It covers the area 'from gene to plant' and aims to enhance genetic diversity and develop improved rice varieties with consumer preferences and tolerances/resistances to abiotic and biotic stresses, using conventional breeding, double haploid breeding, marker-assisted selection (MAS) and genomic selection (GS).

The program exploits the rich reservoir of genetic resources present in the indigenous germplasm pool of African rice.*Oryza glaberrima*, its wild relatives – *O. barthii* and *O. longistaminata* –.and.*O. sativa*. The programme aims to deploy modern breeding approaches to shorten breeding cycles and to increase genetic gains in developing products of market demand.

Rice Sector Development Hubs

Rice sector development hubs represent key rice-growing environments and different market opportunities across African countries and are linked to major national and regional rice-development efforts to facilitate broader uptake of rice knowledge, technologies and innovations..These are zones where rice research products from the research programs and the Africa-wide rice Task Forces and local innovations are tested, adapted and integrated across the rice value chain to achieve development outcomes and impact. Task Force activities focus on the rice hubs to avoid dispersion of activities.

Hubs are built around large groups of farmers and involve other value-chain actors, such as rice millers, seed enterprises and input dealers, and rice marketers. Hubs work to improve value chains by investing in institutional innovations and market development. Innovation platform (IP) is a key catalyst of the rice sector development Hubs.

The main objective of the rice hubs is to enhance the productivity and competitiveness of the rice value chain through; (i) Integration of rice systems in partnerships and businesses (ii) Increased adoption of appropriate technologies and innovations on rice (iii) Knowledge sharing and learning (iv) Improved governance structures and systems of IPs in the rice hubs. In total, 74 Hubs in various agro-ecological zones have been identified by national programmes in 25 countries. In 2016, a decision was made to make a difference between 'core' hubs and 'satellite' hubs.

Pan-African Rice Seed Capital

The rice sector in sub-Saharan Africa is plagued by the low availability of quality seeds of improved rice varieties as well as by the absence of appropriate policies to encourage the development of the seed sector. AfricaRice member countries as well as private seed enterprises that are interested in boosting the domestic production of rice have expressed the need for a sustainable access to sufficient amounts of breeder and foundation seeds of the most popular improved rice varieties developed by AfricaRice and its national partners. These improved varieties include ARICAs, NERICAs and recently developed hybrids.

A sustainable breeder and foundation seed capital is essentially required to respond to the continuous and increasing needs for certified seeds of governments, private enterprises, and farmers' cooperatives. It will also serve as a seed stock in case of disasters such as droughts, flooding, and civil wars. The existence of a Pan-African Rice Seed Capital will contribute to the establishment of effective, efficient, and highly performing domestic rice value chains.

A Pan-African Rice Seed Capital has thus been established at the AfricaRice Research Station in M'bé, Côte d'Ivoire, by AfricaRice and its partners. It is based on a public–private partnership business model that the Center has developed to enhance the rice-seed value chain. The model creates synergy between agribusiness and smallholders to meet the seed needs for food security and generate value addition and jobs for youth and women.

The Pan-African Rice Seed Capital will strengthen collaboration between all actors of the rice value chain: AfricaRice, NARIs, private seed enterprises and seed cooperatives, paddy producers, processors, transporters and traders of milled rice.

It will help to establish functional rice seed systems in each of the member countries to address the issue of rice self-sufficiency by providing enough quality seeds. This will facilitate farmers' access to quality seeds in order to increase productivity and the quality of locally-produced rice and will contribute to generating increased revenue from the use of improved varieties (www.AfricaRice.org).

2. International Centre for Agricultural Research in the Dry Areas (ICARDA)

The International Center for Agricultural Research in the Dry Areas (ICARDA), headquartered at Beirut, Lebanon is an international organization undertaking research-for-development. ICARDA provides innovative, science-based solutions for communities across the non-tropical dry areas. In partnership with research institutions, NGOs, governments, and the private sector, ICARDA's work advances scientific knowledge, shapes practices, and informs policy (www.icarda.org).

Since its establishment in 1977, ICARDA has implemented research-for-development programs in 50 countries across the world's dry areas – from Morocco in North Africa to Bangladesh in South Asia.

ICARDA has a global mandate for the crop development of barley, lentil, and faba bean, and serves the non-tropical dry areas for the improvement of water-use efficiency in agriculture, rangeland issues, and small-ruminant production. In the Middle East and North Africa region, and in Central Asia, ICARDA contributes to the improvement of bread and durum wheat, *kabuli* chickpea, pasture, and forage legumes, and associated farming systems. It also works on land management topics, the diversification of production systems, and value chains for sector-based crop and livestock products. Social, economic and policy research, as well as communication and knowledge sharing, is an integral component of ICARDA's approach to enhance the uptake of new technologies and maximize the impact of research outputs.

Mission

To reduce poverty and enhance food, water, and nutritional security and environmental health in the face of global challenges, including climate change.

Vision

ICARDA envisions thriving and resilient livelihoods in the dry areas of the developing world with adequate incomes, secure access to food, markets, and nutrition, and the capacity to manage natural resources in equitable, sustainable, and innovative ways.

Research Programmes

To address critical challenges and strengthen resilience across the dry areas ICARDA advances scientific knowledge, shapes practices, and informs policy through three research programs.

1. Biodiversity and crop improvement
2. Resilient agricultural livelihood systems
3. Water, land and ecosystems

Out of the three research programmes, biodiversity and crop improvement is relevant to the crop breeding and hence the same is described as follows.

Biodiversity and Crop Improvement

ICARDA's Biodiversity and Crop Improvement Research Programme develop resilient plant varieties that can meet current and future challenges faced by the non-tropical dry areas. The programme applies conventional and molecular breeding to develop highly-adapted crops with resistance or tolerance to major biotic and abiotic constraints.

Crop breeding efforts involve: introgressing new alleles from landraces and wild relatives into elite germplasm; mainstreaming nutritional quality (including bio-fortification) into current breeding programmes; and widening the genetic base as a major strategy to realize the full potential of yield, yield stability, quality, and nutrition.

Given that water scarcity is a key driver of yield instability in the dry areas, the Biodiversity and Crop Improvement Program identifies genotypes with better water-use efficiency to minimize yield losses during drought and maximize yield gains during good seasons. .

Gene Bank and International Nurseries

The Programme manages a gene bank containing some 157,000 samples of major spring and winter cereals, food legumes, and forage and rangeland species – many drawn from the 'Fertile Crescent' in Western Asia, the Abyssinian highlands in Ethiopia, and the Nile Valley, where the earliest known crop domestication practices were first recorded. Many plants are now extinct or endangered in their natural habitats.

The Center's crop improvement programme, and its national partners, use the genetic resources maintained in the ICARDA gene bank to develop nurseries for a wide range of agricultural systems, distributing them worldwide upon request. This service, an integral part of an international nursery trialing system, offers cooperators an opportunity to evaluate genetically diverse germplasm generated through conventional and modern breeding methodologies under their own agro-ecological conditions and socio-economic contexts..

Over the past four decades the Programme's improved cereal and legume varieties have been tested and released by national programmes in partnership with ICARDA and adopted by farmers worldwide. It is estimated that these varieties generate net benefits of approximately US$ 850 million each year. The Program contributes to the development of new Open Access tools for electronic data capture, data analysis, and decision support – thereby making data on genetic resources and bio-informatics more widely available; and also assesses the status and threats to dry areas agro-diversity, recommending management plans for the conservation of biodiversity-rich areas..

3. International Crops Research Institute for the Semi-Arid Tropics (ICRISAT)

The International Crops Research Institute for the Semi-Arid Tropics (ICRISAT) is a non-profit, non-political organization that conducts agricultural research for development in the drylands of Asia and sub-Saharan Africa. Covering 6.5 million square kilometers of land in 55 countries, the semi-arid or dryland tropics has over 2 billion people, and 644 million of these are the poorest of the poor.

ICRISAT and its partners help empower these poor people to overcome poverty, hunger and a degraded environment through better agriculture. ICRISAT is headquartered in Hyderabad, Telangana State, in India, with two regional hubs (Nairobi, Kenya and Bamako, Mali) and country offices in Niger, Nigeria, Zimbabwe, Malawi, Ethiopia and Mozambique.

ICRISAT conducts research on five highly nutritious drought-tolerant crops: chickpea, pigeonpea, groundnut, sorghum and pearl millet.

Chickpea

In a scientific advance involving scientists from more than 20 research organizations, sequencing of the chickpea genome was recently completed for 90 chickpea genotypes, including several wild species. The researchers responsible for the work, led by ICRISAT, succeeded in identifying more than 28,000 genes and several million genetic markers that scientists expect will lead to the development of superior varieties.

Chickpeas are susceptible to several major diseases and insect pests and yields can fall precipitously if the crop is exposed to extreme temperatures or drought. At many locations farmers must produce their crops in 90 to 120 days to avoid the risks associated with drought and high temperatures. In response to these problems ICRISAT scientists have developed new, more robust early maturing cultivars with higher levels of disease resistance and excellent drought tolerance.

Modern early maturing chickpea lines have had a considerable impact in short season environments across South and Southeast Asia transforming what was once a subsistence crop to into significant income generator and export commodity.

In Myanmar, the world's sixth largest producing nation, early maturing cultivars now cover some 80% of the area devoted to chickpea. Adoption of improved cultivars has led to a four-fold increase in the production. Indian farmers have reported comparable productivity increases ranging from 50 to 90% percent.

Currently, efforts are being made to enhance tolerance to terminal drought and heat stresses. In addition, researchers are working to boost the crop's natural nutritional advantages by increasing its protein, iron and zinc content. Plans also call for developing varieties that are resistant to herbicides and can be harvested with small machinery to reduce costs and overcome labor shortages that arise at weeding and harvest time. These efforts should help to reduce production shortages, stabilize prices for producers and consumers, and help to drive the development of overseas markets.

Pigeonpea

Orphaned long ago, pigeonpea is making a comeback in the face of drought and climate change. Once considered of little significance, pigeonpea is rapidly gaining a reputation as a food security crop, income generator and commercial export commodity. Pigeonpea is grown by millions of resource-poor farmers on marginal land across the semi-arid regions of Asia and Africa. Producing small seeds similar in appearance to soybean, pigeonpea is valued by farmers and consumers as food and forage crop. It is also a major contributor to food security in areas facing the early effects of global climate change.

Crop protection specialists estimate that the pests and diseases that attack pigeonpea are responsible for US$ 1.1 billion in annual losses across the semi-arid tropics. Principal plant diseases include Fusarium wilt and Sterility mosaic; major insect pests are pod borers, mainly *Helicoverpa, Maruca* and Pod fly.

A major progress was achieved with the development of the world's first commercial pigeonpea hybrid. This hybrid technology was developed after intensive research efforts spanning more than three decades and has provided the opportunity to break a decades-old yield plateau in pigeonpea. This system, in combination with natural out-crossing of the crop, was successfully used to develop economically viable hybrid seed production technology. It has been refined to suit different agro-ecologies and the hybrids ICPH 2671 and ICPH 2740 were released in India for general cultivation by farmers.

Early-maturing pigeonpea varieties in Africa are helping farmers reduce costs and are rapidly gaining access to an expanding market. The new varieties, which are resistant to Fusarium wilt disease, a major production constraint, are being marketed through local agro-dealers. Studies show that they are quickly supplanting small-seeded types that failed to meet basic market requirements. The new cultivars are now widely grown in Malawi, Mozambique, Tanzania and Uganda where they are filling a gap left by the collapse of local bean varieties.

Pigeonpea is the first non-industrialized commodity, and only the second food legume, with a published genome sequence. Genome analysis identified more than 48,000 genes, including several hundred that are unique to the crop and are closely associated with drought tolerance. The availability of the sequence opens up new avenues for crop improvement and could help plant breeders raise yields from an average of 700 kg/ha closer to the crop's full potential, an estimated 3.5 tons/ha. In time it could also lead to the production of short-duration lines that are both temperature and photoperiod insensitive, thereby expanding the geographical range the crop.

Groundnut

At the time of the Spanish and Portuguese conquest, groundnut was widely grown in South America. By the latter part of the 16th century it had spread to West Africa and soon after to South Asia. It is currently grown on about 22 million hectares worldwide, an area roughly equal in size to the United Kingdom.

Groundnut is currently grown on about 21.8 million hectares worldwide, an area similar in size to the United Kingdom.. Global production (2011) totaled 38.6 million tons, 95 percent of which occurred in developing countries. Major producers include China, India, Nigeria, USA, and Myanmar. Common names include earthnuts, goober peas, monkey nuts, peanut, pygmy nuts and pig nuts. Low yields in the semi-arid tropics are attributable mainly to the fact that groundnut is grown in marginal area under rainfed conditions and is subject to periodic drought.

The crop is highly susceptible to contamination by some 20 different mycotoxins, including aflatoxin caused by a fungus *Aspergillus flavus*.. Aflatoxin is extremely hazardous to human health and is especially harmful to the physical and mental development of young children. Over time, exposure to aflatoxin-infected foods can lead to hepatitis, immune system suppression and liver cancer.

The risks associated with aflatoxin contamination have led industrialized countries to establish rigorous quality standards that often deny developing country farmers the opportunity to export. In West Africa, where groundnut is largely produced by women, export prohibitions have important implications for family well-being and frequently prevent farmers from purchasing resources that might otherwise be used to increase productivity.

Although the resources available for groundnut R&D in the semi-arid tropics are modest compared with funding for the major cereal crops, all available resources have been targeted to achieve key objectives for productivity and safety. For example, researchers have developed a simple, inexpensive testing kit to identify the presence of aflatoxin on stored grain. The kits use reagents that are commonly available in developing countries and generate results comparable to the more expensive techniques used in industrialized country laboratories. .They are currently being used in India, Kenya, Malawi, Mali and

Mozambique in an effort to improve consumer safety and help farmers re-enter the European market.

The kits also support national plant breeding programmes and play a central role in the development of new varieties that are now beginning to displace the older cultivars that have long dominated the Indian market.. One such drought-tolerant variety named after the Hindu goddess Devi (ICGV 91114) ranks among India's most popular. Economic projections indicate that annual benefits in just one district where the new varieties are grown should surpass U$ 500 million by 2020.

Sorghum

Sorghum is a highly reliable crop that grows well in hot, dry environments. It is "climate change-ready", and provides food security and income for millions of poor farmers living in such locations..Over half a billion people rely on sorghum as a dietary mainstay and, given its diversity of uses, as an important source of income. The grain is used mainly for food, prepared in the form of flat breads and porridges of different kinds. Sorghum stover is a vital source of fodder for livestock. Sorghum is also used for a wide range of industrial purposes, including the production of sweet syrups, as a source of starch for fermentation, and for producing bio-fuel. The durable stalks of the plant are used by the rural poor as a construction material and as fuel for cooking.

Sorghum is an extraordinarily robust and reliable crop due to following factors.

- It will grow over a wide range of temperatures and elevations;
- It performs well in different soil types – from very porous sandy soils that don't retain water to heavy clay-types that are prone to water logging;
- It has tremendous resilience in the face of drought, and not only survives but still produces grain using only residual moisture;
- It can be grown under a wider range of soil acidity than many other crops; and
- It is resistant to grain mold, giving people and animals protection against the health dangers of contamination by mycotoxins.

These attributes make sorghum a critically important food security crop for millions of people living in harsh, dry environments – places that don't readily support crops such as maize. The cost of producing sorghum and maize are about the same, as is their nutritional value, so farmers often decide which crop to grow based on local environmental conditions. Given that sorghum requires less water, it is usually grown instead of maize in the hotter and drier areas of

Africa, South Asia, and Central America.

Major sorghum production constraints are as follows.

- Shoot fly, stem borer, head bug and aphid insect pests;
- Grain mold, anthracnose diseases, and leaf blight;
- Weed competition and (in Africa) the parasitic plant *Striga* spp.; and
- Abiotic stresses such as drought (especially terminal drought), high temperatures, acid soils (which are often associated with toxic levels of aluminum saturation) and low soil fertility (in terms of both macronutrients such as nitrogen and phosphorus, as well as micronutrients such as iron and zinc).

Current and on-going research by ICRISAT is aimed at increasing yield potential, disease and insect pest resistance, and tolerance to such abiotic stresses as drought, heat, and various soil-related limitations. All this research is done within the context of the Institute's Inclusive Market-Oriented Development (IMOD) framework which seeks to harness markets for the rural poor. Looking ahead, research opportunities include the following.

- Creation of hybrids to increase yields for a wider range of production systems in Africa, building on successes in India, Mali and elsewhere;
- Development of improved "dual purpose" plant types for grain, as well as feed and fodder, that can increase the value of the crop and strengthen the integration of animal husbandry with crop production, resulting in higher and more stable incomes while improving soil health through increased organic matter cycling; and
- Efficient use of the crop's rich genetic diversity for the improvement of sorghum and other cereals, made possible by the availability of the full genome sequence and other genetic and genomic tools that will enable identification and transfer of favorable alleles for stress tolerance (such as phosphorus deficiency, aluminum toxicity and terminal drought), product quality (micronutrient content, digestibility and industrial qualities) and superior agronomic performance.

Over the past four decades, a total of 242 cultivars have been released using germplasm and ICRISAT-bred lines and hybrid parents in all regions (Asia, Eastern and Southern Africa, West and Central Africa, and Latin America). The number of cultivar releases has been highest in Asia (75), including 35 in India, closely followed by Eastern and Southern Africa (74), West and Central Africa (58), and Latin America (35). While released cultivars include both hybrids and varieties in Asia, in Africa releases were composed mostly of varieties (an exception being a hybrid released in Sudan).

Pearl Millet

Pearl millet is a dependable nutritious source of food for millions in marginal agricultural areas. It is the world's hardiest warm-season cereal crop. It can survive and produce grain even on the least fertile soils in the driest regions, on highly acidic and saline soils, and in the hottest climates. More than 90 million desperately poor people depend on pearl millet for food and income. They generally live in the drier parts of Africa and Asia, places where most other crops just won't grow, and local farm households literally have nowhere else to turn for food security. Fortunately, pearl millet is not just a resilient and dependable source of energy, but also a good source for other dietary needs, especially micronutrients.

Pearl millet covers an estimated 31 million hectares worldwide – an area about the size of Poland – and is grown in more than 30 countries located in the arid and semi-arid tropical and subtropical regions of Asia, Africa and Latin America. The crop is well adapted to agricultural areas that are afflicted by severe drought, poor soil fertility, and high temperatures. It also grows well in soils with high salinity or low pH (highly acidic). Because of its tolerance to such challenging environmental conditions, it is often found in truly marginal areas where other.cereal.crops, such as.maize.or wheat, can't survive. It is grown, for example, on the edge of the Sahel desert in Northeastern Mali, an area considered by many to be where the crop came from.

The major constraints to pearl millet production include:

- Such bacterial diseases as bacterial spot (*Pseudomonas syringae*) and bacterial leaf streak (*Xanthomonas campestris*.pv. *pennamericanum*);
- Various fungal and pseudo-fungal diseases, especially downy mildew (caused by *Sclerospora graminicola* and *Plasmopara penniseti*), blast (caused by *Pyricularia grisea*), smut (caused by *Moesziomyces penicillariae*), ergot (caused by *Claviceps fusiformis*) and rust (caused by *Puccinia substriata* var. *penicillariae*);
- The parasitic weed, *Striga hermonthica*, which can significantly reduce productivity;
- Insect pests, including millet head miner and stem borers;
- Parasitic nematodes; and

- Abiotic stresses such as drought, soil acidity, soil salinity, and high temperatures during the time when seedlings are just starting to grow and when the plants are flowering.

ICRISAT scientists have conducted extensive research on resistance to the major insect pests and diseases that limit pearl millet production, as well as on improving the crop's natural tolerance to abiotic stresses, including drought, poor soils and high temperatures. Considerable effort has gone into the development of high yielding, early maturing pearl millet hybrids, especially for the sufficiently moist rainfed areas (600-800 mm of annual precipitation) and irrigated areas of the semi-arid tropics of the Indian Subcontinent. Future research will build on the growing interest in higher yielding hybrids for Sub-Saharan Africa, building on past successes in India and on the initial hybrid vigor classification of pearl millet landraces from West Africa. Research will focus on exploiting the high levels of mineral micronutrients (iron and zinc) found in pearl millet grain, the crop's high optimum temperature for growth, and its increased use for alternative food products, feed, and fodder; and work will also continue to develop new genetic and genomic tools for identifying and deploying favorable genes that can significantly improve grain and stover yield potential, biotic stress resistances and abiotic stress tolerances, as well as the nutritional value of pearl millet grain, green fodder and stover.

Since 1982, the year when WC-C75 – the first ICRISAT-bred open-pollinated variety (OPV) – was released for cultivation in India, 67 cultivars based on ICRISAT-bred germplasm (14 OPVs and 53 hybrids) have been released and notified in India. Five of the OPVs were developed by ICRISAT, and nine were produced by national agricultural research partners. Five of the hybrids were developed by ICRISAT, and 31 by national partners using ICRISAT-bred male-sterile lines.

Seventeen hybrids developed by the private sector are also based on ICRISAT-bred male-sterile lines, or on selections made within these lines. Many other hybrids from the private sector, and several others marketed as truthfully labeled seeds, also involve some degree of ICRISAT-bred materials – both seed parents and pollen parents.

Genebank at ICRISAT

The genetic diversity of the world's major food crops, once considered virtually limitless, is fast disappearing in nature. The causes are well-known: the predominance of modern crop varieties, changes in consumer preferences, and loss of habitat each contribute to a trend that could well undermine the planet's food supply. Dryland areas are especially vulnerable to genetic erosion as they lack resiliency and are now experiencing the early impact of global climate change. These factors – in combination with poverty, human migration and political instability – tend to accelerate the problem.

To safeguard the biodiversity of food crops grown in the semi-arid tropics, ICRISAT maintains one of the world's largest germplasm repositories, a "genebank" containing more than 120,000 accessions. The facility, serves as a world repository for sorghum, pearl millet, chickpea, pigeonpea, groundnut and six of the lesser-known millet species – finger millet, foxtail millet, proso millet, kodo millet, little millet and barnyard millet. Maintained at low temperatures to preserve viability, the collection contains hundreds of wild species as well as many farmer-developed varieties or land races that are fast disappearing in nature.

Located at the Center's headquarters in Patancheru, India, the genebank was designed to withstand natural disasters and has numerous backup systems to keep the collection secure.. In addition, three smaller facilities have been established in Africa to serve national and regional research programmes and provide additional safeguards. .Nearly 100,000 duplicate samples are stored at the Svalbard Global Seed Vault in Norway under the auspices of the Global Diversity Trust and the Nordic Genebank.

The purpose of maintaining a genebank is not to simply safeguard genetic diversity, but to encourage its use. To that end, researchers have characterized nearly 95% of the collection for morpho-agronomic traits, and a greater part for stress tolerance, pest and disease resistance and nutrition. Each accession maintained in the collections is freely available to researchers at other institutions. Thus far, more than 720,000 samples have been distributed to users in 146 countries (www.icrisat.org).

4. International Institute of Tropical Agriculture (IITA)

IITA, headquartered at Ibadan, Nigeria is an award-winning, research-for-development (R4D) organization, providing solutions to hunger, poverty, and the degradation of natural resources in Africa. Since 1967, IITA has worked with international and national partners to improve livelihoods, enhance food and nutrition security, increase employment, and preserve natural resource integrity.

IITA is guided by an ambitious strategy–to lift 11.5 million people out of poverty and revitalize 7.5 million hectares of farmland–by 2020. As one of 15 research centers in the CGIAR, a global partnership for a food secure future, IITA is engaged in several CGIAR Research Programmes (CRPs).

IITA has delivered more than 70% of the CGIAR's impact in sub-Saharan Africa and remains committed to science-driven improvement of agriculture and related food value chains. Its R4D programs are focused on four crucial areas:

- Biotechnology and Genetic Improvement,
- Natural Resource Management,
- Social Science and Agribusiness,
- Plant Production and Plant Health and Nutrition and Food Technology.

IITA's R4D programs have attracted the best and brightest minds from all over the world. Thousands of scientists and professionals have been involved in IITA training and research programmes and continue to benefit from the knowledge they have in turn passed on to others. To help address poverty and the high number of unemployed young men and women in, IITA has initiated a Youth Agripreneurs programme to create jobs by making agriculture and agribusiness appealing to the youth. The programme is integrated into the Business Incubation Platform (BIP), the technology delivery arm of IITA, which serves as a model to stimulate product development and to provide opportunities for market expansion (www.cgiar.org).

IITA focuses on research themes that address staple food crops including banana, plantain, cassava, cowpea, maize, soybean and yam. IITA is well known for cowpea improvement and its germplasm holding. The same is briefly described as follows.

IITA's Cowpea Research and Impact

IITA scientists have developed high yielding varieties that are early or medium maturing and have consumer-preferred traits such as large seeds, seed coat texture and color. A number of the varieties have resistance to some of the major diseases, pests, nematodes, and parasitic weeds. They are also well-adapted to sole or intercropping.

Following more than two decades of research, field trials, and risk assessment by multiple organizations, on 29 January 2019, the Nigerian Biosafety Management Agency (NBMA) approved the commercial release of genetically modified cowpea to farmers in Nigeria. NBMA's approval allows the Institute for Agricultural Research (IAR) to commercially release Pod Borer-Resistant Cowpea (PBR Cowpea)-event AAT709A, genetically improved to resist the pod borer (*Maruca vitrata*).

Improved varieties have been released to 68 countries in all of the world's regions. Additionally, IITA's Farmer Field School (FFS) projects, in collaboration with partners, have trained farmers in improved pest management practices of cowpea crops.

The IITA genebank holds the world's largest and most diverse collection of cowpeas, with 15,122 unique samples from 88 countries, representing 70% of African cultivars and nearly half of the global diversity (www.iita.org).

Genebank at IITA

Preserving biodiversity outside of its natural environment is done in the Genebank. IITA's Genebank holds over 28,000 accessions of plant material or germplasm, of major African food crops. This germplasm is held in trust on behalf of humanity under the sponsorship of the United Nations. It is distributed without restriction for use in research for food and agriculture. Started in the mid-seventies the Genebank helps in crop improvement and also provides "seeds of hope" for people affected by flood, fire, wars, and other disasters.

The main crops stored in the Genebank are cowpea, cassava, plantain and banana, yam, soybean, bambara groundnut, and maize. In addition, substantial collections of wild cowpea relatives and miscellaneous legumes have been collated over the past 30 years. More recently a small collection of African yam bean, an underused legume, has been put together.

By far the most important crop in the Genebank is the cowpea. The IITA Genebank holds the world's largest and most diverse collection of cowpeas with 15,122 unique samples from 88 countries, representing 70% of African cultivars and nearly half of the global diversity. This incredible collection makes IITA integral in the protection of the cowpea species.

Depending on the species and its reproductive and dissemination biology, collections are either stored in the field, or in the seed or in-vitro genebanks. All crops producing orthodox seeds are maintained at optimal water content in medium-term (5 °C, 30–35% relative humidity) or long-term (-20 °C, under vacuum) storage conditions.

Clonal crops, those that are propagated through cuttings, are either maintained in the field or the in-vitro genebank. Sometimes they are kept in both to maintain the collection

International Maize and Wheat Improvement Centre (CIMMYT)

The.International Maize and Wheat Improvement Center, known by its Spanish acronym, CIMMYT, is a non-profit research and training organization with more than 400 partners in over 100 countries. CIMMYT is the global leader in publicly-funded maize and wheat research and related farming systems. Headquartered near Mexico City, the CIMMYT works with its partners throughout the developing world to sustainably increase the productivity of maize and wheat cropping systems, and thus improves global food security and reducing poverty.

For this it applies the best of agronomy practices and breeding, socioeconomics, agricultural extension, and capacity building to create sustainable solutions with lasting impact and a strong focus on climate change, hunger and nutrition, rural community development, and the environment. CIMMYT's germplasm bank is home to humanity's largest collection of maize and wheat varieties made freely available to scientists, researchers and farmers around the world.

Research themes at CIMMYT are as under.

- Capacity building
- Food security
- Gender and social inclusion
- Health and nutrition
- Innovation and technology
- Climate change

A few important themes/developments directly related to maize and wheat improvement are as follows.

Maize Research

CIMMYT develops and distributes improved maize inbred lines and hybrids to partners worldwide. This work improves the food security, incomes and livelihoods of millions of smallholder farmers and their families in the tropics and subtropics of Africa, Asia, and Latin America.

The objectives of CIMMYT's maize breeding and seed systems work are to:

- Develop genetically diverse, high-yielding maize inbred lines and hybrids, combining multiple-stress tolerance, nutrient use efficiency, enhanced nutritional quality, and desirable seed production properties for target production zones.

- Accelerate the rate at which the genetic potential for maize yields increases each year and improve the efficiency of maize breeding operations, through integrated application of modern tools and technologies.
- Deploy CIMMYT-derived improved maize varieties in the tropics through public-private partnerships, while strengthening local maize seed production and delivery systems.
- Build local capacity — particularly of women and youth — in maize breeding and seed systems through training and knowledge exchange.

CIMMYT's maize breeding work aims at developing easy-to-produce, best-bet hybrids, elite maize lines, and improved open-pollinated varieties, as well as science-based recommendations for varietal targeting and improved productivity in target regions.

In addition to higher grain yield, CIMMYT's maize breeding teams focus on:

- Yield stability in stress-prone environments, through tolerance to drought, heat, and poorly fertile soils.
- Resistance to major diseases, insect pests, and parasitic weeds.
- Enhanced nutritional qualities, such as provitamin-A, high-kernel zinc, and quality protein maize.

This work is supported by cutting-edge tools and technologies, including doubled haploids, molecular markers and high-throughput field-based phenotyping. CIMMYT's research and work on maize breeding and seed systems is accomplished through strong public-private partnerships.

CIMMYT leads the CGIAR Research Programme on Maize.

Research project under maize improvement are as follows.

- Accelerating Genetic Gains in Maize and Wheat (AGG)
- Bio-fortified Maize for Improved Human Nutrition
- Climate Resilient Maize for Asia (CRMA)
- Double Haploid Production Services
- Fall Armyworm R4D and Management
- Heat Stress Tolerant Maize for Asia (HTMA)
- International Maize Improvement Consortium for Africa (IMIC-Africa)
- International Maize Improvement Consortium for Asia (IMIC-Asia)
- Stress Tolerant Maize for Africa (STMA)

Wheat Research

Demand for wheat by 2050 is predicted to increase by 50 percent from today's levels. Meanwhile, the crop is at risk from new and more aggressive pests and diseases, diminishing water resources, limited available land and unstable weather conditions — heat in particular.

CIMMYT's Global Wheat Programme is one of the most important public sources of high yielding, nutritious, disease- and climate-resilient wheat varieties for Africa, Asia, and Latin America. CIMMYT breeding lines can be found in varieties sown on more than 60 million hectares worldwide.

Through this program, CIMMYT works with more than 200 research and breeding institutions, including the International Center for Agriculture Research in the Dry Areas (ICARDA) and the CGIAR Research Programme on Wheat, sharing elite breeding lines and associated data through its system of international nurseries.

The Wheat Molecular Breeding laboratory develops tools and information for breeders around the world. The Wheat Quality laboratory ensures that CIMMYT varieties meet market demands for flour and bread quality.

CIMMYT's wheat research aims to:

- Develop climate resilient, nutritious, high yielding disease and pest tolerant wheat lines.
- Use the latest molecular breeding tools, bioinformatics and selection methods.
- Ensure that national agricultural research system partners are active participants in breeding.
- Apply more precise phenotyping approaches — phenotyping platforms — and other tools, like remote sensing, to develop genetically diverse wheat varieties so that globally, annual genetic yield gains of at least 0.7 percent are achieved.
- Provide diverse, high-yielding wheat varieties that withstand infertile soils, drought, pests and diseases.
- Conduct research to help farmers exploit the full potential of improved seed while conserving soil and water resources.
- Explore new market opportunities for small holder farmers.
- Provide training opportunities in wheat breeding and crop management research.
- Exploit genetic variation in wheat wild relatives.

CIMMYT leads the CGIAR Research Programme on Wheat.

Wheat Projects

- Accelerating Genetic Gains in Maize and Wheat (AGG)
- Borlaug Institute for South Asia (BISA)
- International Wheat Yield Partnership (IWYP)
- International Winter Wheat Improvement Programme (IWWIP)

Genetic Resources

CIMMYT manages humankind's most diverse maize and wheat collections. It starts with the seed. CIMMYT's germplasm bank, also known as a seed bank, is at the center of CIMMYT's crop-breeding research. This remarkable, living catalog of genetic diversity comprises of over 28,000 unique seed collections of maize and 150,000 of wheat.

From its breeding programs, CIMMYT sends half a million seed packages to 800 partners in 100 countries each year. With researchers and farmers, the center also develops and promotes more productive and precise maize and wheat farming methods and tools that save money and resources such as soil, water, and fertilizer.

Related research, conservation, and utilization are supported by three units: Seed Health, Biometrics & Statistics, and Data Management. Finally, the Seeds of Discovery (SeeD) project analyzes and documents maize and wheat biodiversity and facilitates its use in crop breeding to address current and future production challenges (www.cimmyt.org).

CGIAR Breeding Programmes Need More Than Just Tech Upgrades — They Need Change Management

Here are four ways plant breeding programmes can ready themselves for the big changes that needs to make.

The Case for Change in Plant Breeding Programmes

Plant breeders are in fact missing some vital opportunities. For example, there continues to be a rather limited use of real market insights to inform resource allocation within programmes. This in turn results in a selection of traits weighted towards what breeders and associated scientists think are needed, which may not necessarily meet actual market needs. With new goals and structures foisting change on breeding programs, their success depends on one thing above all else: savvy change management. Fortunately, there are some steps CIMMYT scientists can take to manage change well.

1. Drive Out Complacency With a Sense of Urgency

Most change management efforts fail when insufficient urgency is built early enough in the process. But this urgency can be the most effective antidote against complacency. Organizations that have either secured a very dominant and successful position in the market, or lack effective and threatening competition, can very easily slide into a sense of self-righteousness and an inward-looking perspective.

Although CGIAR breeding efforts could be thought of as an example of the latter — lacking competitors — seasoned managers in industry and marketing like to think that "there is no such thing as a lack of competition." Funding, for one, is by nature a competition. Funding agencies might look at other fields and/or players to support if they deliver a higher return on investment, not only financially but also socially.

The impact of high complacency cultures can be seen in plant breeding. For instance, a rather large number of breeding programmes still lack a high enough rate of what is called "elite x elite crosses." Unless breeding pipelines run on such crosses, they achieve less than optimal genetic gains and delivery at the field. And donors get a lesser return on investment. Moreover, this complacency means not delivering the best varieties smallholder farmers need to support their families.

2. Build a Guiding Coalition

These days, driving change is too complex to be led by single individuals. We live in fast-paced times. And situations are full of evident and not-so-evident links among myriad moving pieces. We cannot expect one individual to be able to gather enough information fast enough, and then to consistently make the right decisions. Instead, a guiding coalition is needed, with sufficient determination, commitment and thought diversity. Such coalitions require five traits: a position of power, credibility, leadership, expertise, and individual egos held at bay. Once such teams are assembled, the main drivers of success are having a common goal, and enough trust and safety so the real issues are unearthed and addressed.

3. Develop a Vision and a Strategy

When leadership tries to drive change by applying dated approaches such as micromanagement or an authoritarian stance, plans are likely to fail upon arrival. These methods may breed compliance, but certainly not a fierce and sincere commitment. Because of the extreme uncertainty and organizational survival being at stake, crafting a vision plays a bigger role during change management than during business as usual. Two main aspects of developing a vision are especially relevant to CGIAR breeding programs.

Firstly, academic and R&D organizations often keep doing what has worked well in the past. But any change management effort ought to be very explicit about what it is known as "strategic dismissal." This is the ability to stop and phase out activities no longer providing enough value, or where the outcomes of which are not wanted/needed by funding agencies or beneficiaries. For instance, programmes investing in developing hybrid cultivars for the first time in a crop could downsize previous cultivar development efforts. Alternatively, they could scale down efforts in countries that have their own strong local breeding programmes. These changes are no small feat, but the inability to phase out activities clashes with the very first posit of any effective strategy: don't just "keep doing."

Secondly, a vision provides an invisible fabric that pulls all efforts together in a cohesive way. Therefore, its scope is much wider than most people realize, stretching across strategies, plans, and the budgets and means needed to exert change at the depth and speed needed.

4. Encourage Constructive Confrontation

One characteristic of a complacent organization stands out: a rather low-candor, low-confrontation culture. No one needs excessively high-confrontation, "take no prisoners", toxic cultures. But low-confrontation cultures tend to breed under-performance, status quo maintenance and deeply ingrained complacency. And perhaps the most negative consequence is that they fail to instill a strong enough sense of ownership and accountability among its members.

Retrospective Quantitative Genetic Analysis and Genomic Prediction of Global Wheat Yields

The process for breeding for grain yield in bread wheat at the International Maize and Wheat Improvement Center (CIMMYT) involves three-stage testing at an experimental station in the desert environment of Ciudad Obregón, in Mexico's Yaqui Valley. Because the conditions in Obregón are extremely favorable, CIMMYT wheat breeders are able to replicate growing environments all over the world and test the yield potential and climate-resilience of wheat varieties for all major global wheat growing areas. These replicated test areas in Obregón are known as selection environments (SEs).

This process has its roots in the innovative work of wheat breeder and Nobel Prize winner Norman Borlaug, more than 50 years ago. Wheat scientists at CIMMYT, led by wheat breeder Philomin Juliana, wanted to see if it remained effective. The scientists conducted a large quantitative genetics study comparing the grain yield performance of lines in the Obregón SEs with that of lines in target growing sites throughout the world. They based their comparison on data from two major wheat trials: the South Asia Bread Wheat Genomic Prediction Yield Trials in India, Pakistan and Bangladesh initiated by the U.S. Agency for International Development Feed the Future initiative and the global testing environments of the Elite Spring Wheat Yield Trials.

The findings, published in Retrospective Quantitative Genetic Analysis and Genomic Prediction of Global Wheat Yields, in Frontiers in Plant Science, found that the Obregón yield testing process in different SEs is very efficient in developing high-yielding and resilient wheat lines for target sites. It was found higher average heritabilities, or trait variations due to genetic differences, for grain yield in the Obregón SEs than in the target sites (44.2 and 92.3% higher for the South Asia and global trials, respectively), indicating greater precision in the SE trials than those in the target sites. It was also observed significant genetic correlations between one or more SEs in Obregón and all five South Asian sites, as well as with the majority (65.1%) of the Elite Spring Wheat Yield Trial sites. Lastly, it was seen high ratio of selection response by selecting for grain yield in the SEs of Obregón than directly in the target sites.

The results of this study make it evident that the rigorous multi-year yield testing in Obregón environments has helped to develop wheat lines that have wide-adaptability across diverse geographical locations and resilience to environmental variations. This is particularly important for smallholder farmers in developing countries growing wheat on less than 2 hectares who cannot afford crop losses due to year-to-year environmental changes.

In addition to these comparisons, the scientists conducted genomic prediction for grain yield in the target sites, based on the performance of the same lines in the SEs of Obregón. They found high year-to-year variations in grain yield predictabilities, highlighting the importance of multi-environment testing across time and space to stave off the environment-induced uncertainties in wheat yields. While these results demonstrate the challenges involved in genomic prediction of grain yield in future unknown environments, it also opens up new horizons for further exciting research on designing genomic selection-driven breeding for wheat grain yield.

This type of quantitative genetics analysis using multi-year and multi-site grain yield data is one of the first steps to assessing the effectiveness of CIMMYT's current grain yield testing and making recommendations for improvement—a key objective of the new Accelerating Genetic Gains in Maize and Wheat for Improved Livelihoods (AGG) project, which aims to accelerate the breeding progress by optimizing current breeding schemes.

International Potato Centre (CIP)

With more than 45 years of research-for-development work in potato and sweet potato, the International Potato Center (CIP) headquartered at Lima, Peru has contributed to greater food and nutrition security, economic growth and prosperity. CIP breeders and plant scientists work with local partners and farmers to develop and manage potato and sweet potato varieties that are more resilient to the extremes of climate, pests and diseases, produce higher yields, and have better nutritional and culinary qualities.

The centre's social, nutrition and food scientists bring the same dedication to helping rural farmers and communities understand, adopt, and profit from agricultural and postharvest technologies and best practices tailored for different agro-ecologies, production systems and value chains. In partnership with governments, businesses and international organizations, CIP is scaling these science innovations and approaches, putting the tools for better harvests, incomes and health into the hands of millions of farmers, processors, traders and their communities.

In 2018, an estimated 821 million people around the world were chronically undernourished. Farmers in the world's poorest regions struggle to produce enough food and earn decent incomes due to a lack of high-yielding varieties and quality seed, poor agronomic practices, overdependence on mono-cropping and barriers to market access. Population growth and climate change are making the challenges of nourishing the planet even more daunting.

In response to the agricultural development challenges of the 21st century, CIP has organized its mission and activities around three programs. Two of those programs focus on R&D to enhance the ability of sweet potato and potato to improve the food and nutrition security, livelihoods and climate resilience of smallholder farmers and other value chain actors in different agro-ecologies worldwide. The third program focuses on the CIP genebank, which safeguards collections of sweet potato, potato and Andean root and tuber crop diversity that form the foundation of CIP's R&D operations and results-oriented initiatives.

CIP also leads the CGIAR Research Program on Roots, Tubers and Bananas (RTB), a collaboration of various research centers that seeks to harness the untapped potential of potato, sweet potato, banana, cassava and yam to improve the food security, nutrition, income, climate change resilience and gender equity of smallholders in the developing world. The cross-crop learning and partnerships facilitated by RTB expand CIP's scope and effectiveness.

Crops

Potato, sweet potato and Andean roots and tubers

Programmes

- Potato agri-food systems
- Sweet potato agri-food systems
- Biodiversity for the future

Strategic Programme Objective

Early-maturing, market-preferred, and bio-fortified potatoes and high-quality seed potato will improve productivity and farm incomes of more than 7 million households in Africa, Asia and Latin America by 2023.

Breeding Better Potatoes

Recognizing the urgent need for more nutritious, more resilient food crops, CIP breeders are harnessing the latest scientific knowledge and tools to develop better varieties: early-maturing, stress-tolerant, disease-resistant potato varieties with characteristics desired by consumers and processors. Early maturing potatoes allow the crop to be grown during fallow periods of cereal-based systems. This relieves pressure on scarce land and water resources, helps to increase economic and nutritional value, and eases the strain of food price inflation. Stress-tolerant and disease-resistant potato varieties enable farmers to cope with problems that are expected to grow worse under climate change. They also reduce the need to use agrochemicals which saves money and reduces environmental impacts. CIP has also made significant advances in developing bio-fortified potatoes. These combine elevated levels of iron and zinc with resilience traits to contribute to global efforts to end malnutrition.

Genetic Diversity: Conservation, management and DNA sequencing.

Breeding: Nutritional improvement, increased yields; climate resilience; participatory varietal selection.

Seed System Development: Seed production, multiplication, distribution and quality control.

Sustainable Intensification: Nutritional education, agronomic training and post-harvest handling and storage.

Value Chains: Business training, improved product processing, enhanced market linkages.

Inclusion: Technologies and methods targeted particularly at women and young people.

Resilient Bio-fortified Potato

This product plays a central role in CIP efforts to reduce malnutrition and increase the sustainable intensification, diversification and resilience of agri-food systems. Using locally-adapted iron- and zinc-rich potato varieties, scientists developed a breeding population of potatoes which were subsequently crossed with advanced clones to create improved varieties that are highly productive, bio-fortified, stress-tolerant, disease resistant, and early maturing. In addition to assessing the nutritional benefits of these potatoes, scientists—using participatory, gender-responsive breeding methods—are evaluating them for possible release as national varieties. They will use them to breed regionally adapted candidate varieties for targeted agro-ecosystems in Asia, Africa and Latin America. In partnership with advanced public and private organizations, CIP is also unlocking potato genomics to enhance and accelerate this process in response to agro-ecological conditions and consumer demand.

Biotech Disease Resistant Potato

Late blight is a potentially devastating potato disease globally. CIP has made significant advances in the use of advanced biotechnology and molecular methods to transfer late blight-free resistance from crop wild relatives to cultivated potatoes. These varieties were then thoroughly evaluated at the ILRI-BecA (Biosciences eastern and central Africa) Hub in Kenya and the National Agricultural Research Organization in Uganda and were found to be highly resistant to late blight. CIP's next targets are potato virus Y (PVY) and bacterial wilt—the second and third most significant potato diseases. Building on early work, the Ryadg gene, which confers resistance to all known potato virus Y strains is being cloned for transfer into the late blight-resistant potato. Innovative research is needed to understand bacterial wilt pathogenicity and virulence before durable resistance can be developed to that disease.

Seed of Change-Hybrid Potatoes

Drawing on the remarkable potato biodiversity safeguarded in CIP's genebank, this product concentrates on production of hybrid true potato (botanical) seed, via the cultivation of inbred lines for the capture of hybrid vigor (heterosis).

This is a game-changing breeding strategy; the botanical seed multiplication rate is nearly 150 times greater than that of seed tubers and it facilitates the faster inclusion of new trait combinations. Forecasts of adoption and impact modeling will guide development and establishment of private- public partnerships to ensure targeted farmers are reached.

Biodiversity for Future

The genebank drives the efforts of the International Potato Center (CIP) to conserve the world's genetic diversity —cultivated, wild and breeding material—of potato and sweet potato for future use. It plays a critical role in facilitating the impact-oriented release of CIP innovations and products, particularly suitable varieties for farmers and consumers. In situ and ex situ conservation of genetic diversity is critical for preserving and monitoring changes in the world's plant genetic resources for food and agriculture. Lost genetic diversity—particularly of crop wild relatives—would restrict the ability of crop breeders and researchers to enhance farmers' resilience and ability to produce enough nutritious food for the world.

The CIP genebank conserves—in vitro and in seed— the world's most extensive collections of potato, sweet potato and their wild relatives, as well as an unique collection of Andean roots and tubers—the genetic, physiological and biochemical attributes of which the scientific community has just begun to explore. CIP safeguards that biodiversity in trust for humanity to ensure its availability for breeding and other uses now and in the future. CIP employs cryopreservation of living plant material at -196°C and backs up its seed collection at the Svalbard Global Seed Vault in Norway, as well maintains a vast herbarium collection supporting scientific research. CIP also works closely with Andean communities on in situ conservation of potato diversity and have repatriated thousands of accessions previously lost to them due to civil unrest, disease or climate change.

The genebank serves as a model through its advanced research, public database and interactive use of collections. CIP works with other genebanks to ensure that clean material from its collections is backed-up, preventing the loss of diversity already in conservation. Safeguarding crop biodiversity and enhancing the efficiency of genetic resources conservation play critical roles in facilitating the development and release of new varieties for farmers and consumers across the globe (www.cipotato.org).

International Rice Research Institute (IRRI)

Rice is the daily staple for over 3.5 billion people, more than half of the global population and the majority of the world's poor. Challenges such as increasing populations, dwindling natural resources, and the adverse effects of climate change are threatening the world's rice supply and leading to food and nutrition insecurity.

The International Rice Research Institute (IRRI) headquartered at Los Banos, Manila, Philippines is the world's premier research organization dedicated to reducing poverty, hunger, and malnutrition through rice science. IRRI aims to improve the health and welfare of those who depend on rice-based agri-food systems, and promote and protect the environmental sustainability of rice farming for future generations.

IRRI is an independent, nonprofit, research and educational institute, founded in 1960 by the Ford and Rockefeller foundations with support from the Philippine government. The institute has offices in 17 rice-growing countries in Asia and Africa.

Goal 1: Innovation Leadership for the Global Rice Sector

Working with advanced research institutes and national partners around the globe, IRRI will discover, translate, and integrate deep scientific advancements that enable the adoption of technologies, practices, and policies to solve complex global problems and serve our beneficiaries in rice-growing countries and beyond.

Goal 2: Catalyze Impact at Scale for People and Planet

Across rice-growing countries in rural and urban communities, IRRI will speed up the translation of targeted innovations into the local rice value chain through partnerships, education, and technology to facilitate appropriate adoption, maximize impact in the shortest time, and produce substantive benefits for rice farmers, producers, and consumers.

Goal 3: Transform Rice-based Agri-food Systems

Capitalizing on its scientific and capacity building prominence and track record of delivery, IRRI will purposefully engage with global actors to inform policies and establish standards and benchmarks that transform how food is cultivated, produced, and marketed in rice-based agri-food systems.

To achieve its goals, IRRI works towards 5 Impact Challenges, which guide IRRI efforts and map the pathways of our research.

- Climate Change and Sustainability
- Nutrition and Food Security
- Environment
- Prosperity
- Social Equity

IRRI is the lead institute for the CGIAR Research Programme on Rice (RICE CRP), and is working towards the Sustainable Development Goals-SDGs (www.cgiar.org).

Research Activities

Hybridization

Planting of parental lines for crossing work, producing F_1 (1st filial generation) naked seeds and self to F_2 (2nd filial generation) seeds as the end product.

Rapid Generation Advance

Advancement of F_2 seeds to F_6 (6th filial generation) seeds where lines are fixed through single seed descent (SSD) method in 1.5 years.

Biotic Stress Resistance Evaluation Center

High throughput evaluation of rice breeding materials for host plant resistance against various rice diseases and pests (Rice Tungro Virus, Green Leaf Hopper, Brown Plant Hopper, Bacterial Leaf Blight, Rice Blast).

Seed Multiplication

Seed increase of elite breeding lines for further yield trials and those released as varieties by NSIC for commercial planting and seed distribution.

IRRI Genebank

The International Rice Genebank, maintained by IRRI, holds more than 132,000 available accessions as of December 2019. This includes cultivated species of rice, wild relatives and species from related genera. The genebank is the biggest collection of rice genetic diversity in the world. Countries from all over the world send their rice to IRRI for safe keeping, and for sharing for the common public good.

Traditional varieties and the wild species of rice are being lost through genetic erosion. Farmers adopt new varieties, and cease growing the varieties that they have nurtured for generations and eventually lose these varieties. Wild species are threatened with extinction as their habitats are destroyed by human disturbance. Future crop improvement needs the genetic variation from traditional varieties and related wild species to cope with the many biotic and abiotic stresses that challenge rice production around the world.

IRRI works to ensure the long-term preservation of rice biodiversity as part of the global strategy for the conservation of rice genetic resources. This is in partnership with national programs and regional and international organizations worldwide, including the.CGIAR Research Programme for Managing and Sustaining Crop Collections (www.irri.org).

Each type of rice in the International Rice Genebank is stored in both the base (-20 degrees Celsius, long-term storage) and active (2-4 degrees Celsius, for distribution) collections.

The viability of rice seed can decline very quickly (over months) when stored in conditions more typical of how a rice farmer in Asia may store seed, depending on climate and initial seed quality. For example, if the initial viability of a rice seed sample is 95% (that is, 95% of the seed will germinate) and it is kept at a temperature of 30 degrees Celsius, with 70% relative humidity, and with a seed moisture content of 13.5% (fresh weight basis), seed viability would be 1% after just 5 months.

In the International Rice Genebank, rice seeds have the potential to remain viable for many decades. Viability testing of all samples is carried out every 5 years for seed stored in the active collection, or every 10 years for seed stored in the base collection. If viability falls below 85%, a sample of the remaining seeds is planted to produce fresh seeds for storage.

IRRI continually assesses management procedures in the International Rice Genebank to ensure that the best practices are employed to conserve this vital genetic resource for future generations. Following extensive negotiation among all the contributing countries, IRRI now manages the rice collection stored in the International Rice Genebank under the.International Treaty on Plant Genetic Resources for Food and Agriculture. IRRI supplies free samples of different types of rice seed to any prospective user on request, according to the conditions of the Treaty.

International Centre for Tropical Agriculture (CIAT)

CIAT headquartered at Cali, Colombia works with hundreds of partners to help developing countries make farming more competitive, profitable, and resilient through smarter, more sustainable natural resource management. It helps policymakers, scientists, and farmers respond to some of the most pressing challenges of the present time, including food insecurity and malnutrition, climate change, and environmental degradation. CIAT's global research contributes to several of the United Nations' Sustainable Development Goals, and cuts across four key themes: big data, climate-smart agriculture, ecosystem action, and sustainable food systems.

Research Areas

CIAT (International Center for Tropical Agriculture) develops more resilient, productive, and profitable varieties of cassava and common bean, together with improved tropical forages for livestock. In Latin America and the Caribbean, CIAT also works to boost rice.production and the competitiveness of the region's rice sector. These crops are vital for global food and nutrition security.

The Center's Bioscience Platform includes the largest genebank in Latin America and advanced laboratories where scientists work to accelerate crop improvement.

CIAT also works with partners across Latin America to uphold international standards on biotechnology, and help train the next generation of scientists.

Research on Agro-biodiversity

Standard crop varieties do not always rise to the challenges of modern agriculture: climate change, nutrition and higher productivity on farms. CIAT develops varieties of common bean, cassava and tropical forages (the plants livestock eat) that are often superior to business-as-usual varieties. Because of the 68,000 samples stored in CIAT's seed bank, scientists around the world accelerate crop development to improve livelihoods, diets and farm resilience to climate change.

Bean Research

CIAT safeguards the world's largest and most diverse collection of bean germplasm. The collection consists of around 36,000 samples of cultivated materials mostly from the crop's Mesoamerican and Andean centers of origin together with wild species related to these materials. The CIAT collection constitutes a valuable resource for bean improvement worldwide.

CIAT bean researchers are developing a wide range of user-friendly molecular markers (especially SSRs) and have implemented marker-assisted selection for disease resistance. They also aim to combine tolerance to various abiotic stresses (drought and poor soils) through the use of gene discovery, marker-assisted selection, and farmer participatory methods.

Over 550 new bean varieties released by PABRA member countries with CIAT assistance across Africa since 1996.

Climbing beans yielding three times more than the familiar bush type provide an especially eco-efficient solution for densely populated, land-scarce places like Rwanda, Burundi, and western Kenya (ciat.cgiar.org). CIAT is working on new website for the Alliance of Biodiversity International and CIAT.

11

Commercialized GM Crops

Biotechnology is defined as a set of tools that uses living organisms or parts of living organisms to make or modify a product, improve plants, trees or animals or develop microorganism for specific uses. Agricultural biotechnology is the term used in crops and livestock improvement through biotechnology tools. Biotechnology encompasses a number of tools and elements of conventional plant breeding techniques, bioinformatics, microbiology, molecular genetics, biochemistry, plant physiology and molecular biology. Biotechnology tools that are important for agricultural biotechnology include:

- Conventional plant breeding
- Tissue culture and micro-propagation
- Molecular breeding or marker assisted selection
- Genetic engineering and GM crops
- Molecular diagnostic tools

Transgenics

Transgenesis refers to the process of introducing an exogenous or modified gene (transgene) into a recipient organism of the same or different species from which the gene is derived and the resultant product is known as transgenic or genetically modified organism (GMO). The term transgenic is applied to the organisms that have been altered by introducing DNA molecules into them. It denotes an organism that has been transformed with a foreign DNA sequence. It means that one or more DNA sequences from another species have been introduced by artificial means. Transgenics or GMOs are defined as those organisms with a gene or genetic construct of interest that has been introduced by molecular or recombinant DNA techniques. These exclude organisms produced by conventional breeding as well as organisms produced by the intra-organism rearrangement of genetic materials by physical or by chemical means. The GMO, thus carries "transgene(s)" which when integrated and expressed stably and properly, confer either a new trait to the organism, which was hitherto not present (such as insect resistance), or enhance an already existing trait (such as nutritional quality). Thus, transgenic is an organism that contains a transgene introduced by technological (not breeding) methods and the term transgene means the genetically engineered gene added to a species.

Genetic Engineering and GM Crops

Over the last 30 years field of agricultural biotechnology has expanded rapidly due to greater understanding of DNA as the chemical double helix code from which genes are made. Genetic engineering is one of the modern biotechnology tools that is based on recombinant DNA technology. The term genetic engineering, often interchanged with terms such as gene technology, genetic modification, or gene manipulation, is used to describe the process by which the genetic makeup of an organism can be altered using "recombinant DNA technology." This involves the use of laboratory tools and specific enzymes to cut out, insert, and alter pieces of DNA that contain one or more genes of interest. The ability to manipulate individual genes and to transfer genes between species that would not readily cross is what distinguishes genetic engineering from traditional plant breeding. With conventional plant breeding, there is little or no guarantee of obtaining any particular gene combination from the millions of crosses generated. Undesirable genes can be transferred along with desirable genes or while one desirable gene is gained, another is lost because the genes of both parents are mixed together and re-assorted more or less randomly in the offspring during segregation. These problems limit the improvements that plant breeders can achieve, spending more time and funds along the way

In contrast, genetic engineering allows the direct transfer of one or just a few genes, between either closely or distantly related organisms. Not all genetic engineering techniques involve inserting DNA from other organisms. Plants may also be modified by removing or switching off particular genes and genetic controls (promoters). Genetic engineering techniques are to be used when all other techniques have been exhausted and when: 1) the trait to be introduced is not present in the germplasm of the crop; 2) the trait is very difficult to improve by conventional breeding methods; and 3) it will take a very long time to introduce and/or improve such trait in the crop by conventional breeding methods. Modern plant breeding is a multi-disciplinary and coordinated process where a large number of tools and elements of conventional breeding techniques, bioinformatics, biochemistry, molecular genetics, molecular biology and genetic engineering are utilized in an integrated fashion.

Development of Transgenic Crops

Although there are many diverse and complex techniques involved in genetic engineering, its basic principles are reasonably simple. It is however, very important to know the biochemical and physiological mechanisms of action, regulation of gene expression and safety of gene and gene product to be utilized. The general process of genetic engineering requires the successful completion of a series of six steps (Crop Biotech Updates, 2014).

Step 1. Nucleic Acid (DNA/RNA) Extraction

DNA or ribonucleic acid (RNA) extraction is the first step in the genetic engineering process. It is therefore important that reliable methods are available for isolating these components from the cell. In any isolation procedure, the initial step is the disruption of the cell of the desired organism, which may be viral, bacterial or plant cells, in order to extract the nucleic acid. After a series of chemical and biochemical steps, the extracted nucleic acid can be precipitated to form thread-like pellets of DNA/RNA.

Step 2. Gene Cloning

There are basically four stages in any gene cloning experiment: generation of DNA fragments, joining to a vector, propagation in a host cell, and selection of the required sequence. In DNA extraction, all DNA from the desired organism is extracted. This genomic DNA is treated with specific enzymes called restriction enzymes cutting it into smaller fragments with defined ends to allow it to be cloned into bacterial vectors. Copies of the bacterial vector will then harbor many different inserts of the genome. These vectors are transformed into bacterial cells and thousands of copies are produced. Using information relating to specific molecular marker sequences and the desired phenotype, the vector harboring the desired sequence is detected, selected, isolated and clones are produced. Restriction enzymes are again utilized to determine if the desired gene insert was cloned completely and correctly.

Step 3. Gene Design and Packaging

Once the gene of interest has been cloned, it has to be linked to pieces of DNA that will control its expression inside the plant cell. These pieces of DNA will switch on (promoter) and switch off (terminator) the expression of the gene inserted. Gene designing/packaging can be done by replacing an existing promoter with a new one, incorporating a selectable marker gene and reporter gene, adding gene enhancer fragments, introns, and organelle-localizing sequences, among others.

Promoters

Promoters allow differential expression of genes. For instance some promoters cause the inserted genes to be expressed all the time, in all parts of the plant (constitutive expression) whereas others allow expression only at certain stages of plant growth, in certain plant tissues, or in response to external environmental signals. The amount of the gene product to be expressed is also controlled by the promoter. Some promoters are weak, whereas others are strong. Controlling the gene expression is an advantage in developing GM plants.

Selectable Marker Genes

Selectable marker genes are usually linked to the gene of interest to facilitate its detection once inside the plant tissues. This enables the selection of cells that have been successfully incorporated with the gene of interest, thus saving considerable expense and effort. Genetic engineers used antibiotic resistance and herbicide resistance marker genes to detect cells that contain the inserted gene. Cells that survive the addition of marker agents to the growth medium indicate the presence of the inserted gene. Although increase in antibiotic resistance in humans and animals is unlikely to occur using antibiotic resistance marker, genes coding for resistance to non-medically important antibiotics are preferred. In addition, alternative types of marker genes have been developed which are related to plant metabolism such as phosphomannose isomerase, xylose isomerase and others.

Reporter Genes

Reporter genes are cloned into the vector in close proximity to the gene of interest, to facilitate the identification of transformed cells as well as to determine the correct expression of the inserted gene. Several selectable markers or reporter genes are commonly used to monitor the efficiency of gene transfer and expression pattern of the transgene in various parts of the plant. Such markers are highly useful in analyzing the activities of the promoters in transformed tissues. Since gene transfer is never 100 percent effective due to several reasons, the researcher needs to select genetically transformed regenerants from all the other potential regenerants. If the expression of the desired transgene is not sufficient for early selection of cells or individuals from those that are not transformed, it is necessary to transfer selectable marker gene along with the desired gene. The transformed individuals are then selected on the basis of expression of the marker gene.

The most commonly used selectable markers are genes conferring resistance to antibiotics such as kanamycin and hygromycin, herbicides, like phosphoinothricin and glyphosate or drugs (methotrexate). Earlier, neomycin phosphotransferase II (*nptII*) gene conferring resistance to antibiotic kanamycin or geneticin was widely used as a selectable marker. Currently, however, the bar gene of *Strptomyces hygroscopicus* that provides resistance against herbicides such as Basta, Glufosinate and Bialophos is more commonly used as a selectable marker. The beta glucuronidase gene (*gus A* gene) is derived from *E. coli* and has been the most commonly employed reporter gene for rapid evaluation of transformation protocols. The GUS assay is rapid, sensitive and simple to carry out. It acts on a particular substrate producing a blue product, hence making the transformed cells blue. The green fluorescent protein (gfp) which allows transformed cells to glow under a green light; and luciferase gene that allows cells to glow in the dark, are among others.

The selectable marker introduced in a plant along with the transgene continues its expression in the transgenic plants. Constitutive expression of selectable markers, particularly antibiotic resistance genes, in food crops has generated considerable debate and concern over their bio-safety (Yoder and Goldsorough, 1994). However, now this issue has been tackled successfully.

Enhancers

Several genetic sequences can also be cloned in front of the promoter sequences (enhancers) or within the genetic sequence itself (introns, or non-coding sequences) to promote gene expression. An example is the cloning of the cauliflower mosaic virus promoter (CaMV35S) enhancers in front of the plant promoter. Once the gene of interest is packaged together with the promoter, reporter and the marker gene, it is then introduced into a bacterium to allow for the creation of many copies of the gene package. The DNA isolated from the bacterial clones can then be used for plant cell transformation using particle bombardment. If however the use of bacteria *Agrobacterium tumefaciens* is preferred in the plant transformation, the whole gene package should be cloned in between two border sequences (left and right border) of a binary vector plasmid. This will allow processing of the *Agrobacterium* so that only the transfer DNA (T-DNA) will be incorporated into the plant genome.

Step 4. Transformation

The most common methods used to introduce the gene package into the plant cells in a process called transformation or gene insertion, include biolistic transformation using the gene gun and *Agrobacterium*-mediated transformation.

Particle Bombardment

Particle bombardment is a mechanical method of introducing the desired gene. The desired genetic sequence is cloned into a plant DNA vector and introduced into the plant using the gene gun or particle gun. As in the common gun, the gene gun uses minute particles of tungsten or gold as the bullet. These particles are coated with the DNA solution and fired to the plant cells through the force of the Helium gas inside a vacuum-filled chamber. The DNA and the tungsten/ gold particles get inside the cell, and within 12 hours, the inserted DNA gets inside the nucleus and integrated with the plant DNA. The tungsten/ gold particles are sequestered to the vacuole and eliminated later. Transformed cells are cultured in vitro and induced to form small plants (regeneration) that express the inserted gene.

Agrobacterium tumefaciens-mediated Transformation

The "sharing" of DNA among living forms is well documented as a natural phenomenon. For thousands of years, genes have moved from one organism to another. For example, *Agrobacterium tumefaciens*, a soil bacterium known as 'nature's own genetic engineer', has the natural ability to genetically engineer plants. It causes crown gall disease in a wide range of broad-leaved plants, such as apple, pear, peach, cherry, almond, raspberry and roses. The disease gains its name from the large tumor-like swellings (galls) that typically occur at the crown of the plant, just above soil level. Basically, the bacterium transfers part of its DNA to the plant, and this DNA integrates into the plant's genome, causing the production of tumors and associated changes in plant metabolism. Molecular biologists have utilized this biological mechanism to add desired gene into crops. The genes that cause the galls are removed and replaced with genes coding for desirable traits. Plant cells infected with the bacterium will not form galls but produce cells containing the desired gene, which when cultured in a special medium will regenerate into plants and manifest the desired trait. The main goal in any transformation procedure is to introduce the gene of interest into the nucleus of the cell without affecting the cell's ability to survive. If the introduced gene is functional, and the gene product is synthesized, then the plant is said to be transformed. Once the inserted gene is stable, inherited and expressed in subsequent generations, then the plant is considered a transgenic.

Potrykus (1991) and Christou (1996) reviewed in detail the different gene transfer methods and their advantages and shortcomings. *Agrobacterium* mediated transformation of several dicotyledonous plants is considered now a routine. However, monocots are generally considered recalcitrant (in popular terms-tissue culture shy) to *Agrobacterium*. There have been significant breakthroughs in *Agrobacterium* mediated transformation of monocots like rice and maize. Because of low copy insertion, relatively precise mode of gene transfer, high efficiency of transformation, transfer of relatively large DNA segment and relatively less expenses involved, the *Agrobacterium* mediated strategy is more preferred over other methods. In comparison, direct gene transfer techniques like projectile bombardment often results in multiple-copy integration, transgene silencing, reduced fertility and are still genotype-independent in terms of efficiency. The biolistic approach, on the other hand has several advantages. These include (i) effectiveness regardless of species and tissue type, (ii) relative simplicity and rapid protocol, (iii) ability to transform directly totipotent tissues such as pollen, embryo, meristem, etc., and (iv) unique suitability for organellar transformation as reviewed by Prasanna (2001).

Some of the crops, such as cotton, maize, pigeon pea and chickpea were initially found difficult to transform and regenerate, but there have been significant breakthroughs even in these crops. It is important to note that development of efficient transformation methods is frequently not straight-forward as it can be influenced by many variables such as crop, genetic background of the material, the developmental stage and type of explants, the experimental conditions, etc. In view of the concerns expressed regarding the possibility of gene flow from transgenic plants to wild relatives or weeds through pollen, an increasing emphasis has also being paid for organellar transformation (Maliga, 1993). However, now production of successful commercial transgenic plants has become routine especially by the multinational private seed companies with strong R & D.

Step 5. Detection of Inserted Genes

Molecular detection methods have been developed to determine the integrity of the trans-gene (introduced gene) into the plant cell. Polymerase chain reaction or PCR is a quick test to determine if the regenerated transgenic cells or plants contain the gene. It uses a set of primers (DNA fragments) – forward and backward primers, whose nucleotide sequences are based on the sequence of the inserted gene. The primers and single nucleotides are incubated with the single stranded genomic DNA and several cycles of DNA amplification is conducted in a PCR machine. Analysis of the PCR products in agarose gel will show if the plants are really transformed when DNA fragments equivalent in size with the inserted gene is present and amplified.

Southern blot analysis

This technique determines the integrity of the inserted gene: whether the gene is complete and not fragmented, at the correct orientation, and with one copy number. The DNA coding sequence is the probe binding to the single stranded genomic DNA of the transgenic plant which is implanted on a nitrocellulose paper. Autoradiography will reveal the transgenic status of the plant. One copy of the trans-gene is desired for optimum expression.

Northern blot analysis

This technique determines whether the transcript or the messenger RNA (mRNA) of the introduced DNA is present and is correctly transcribed in the transgenic plant. The messenger RNA of the transgenic plants are isolated and processed to bind to the nitrocellulose membrane. Labeled DNA is used to bind to the mRNA and can be visualized through autoradiography.

Western blot analysis or protein immuno blotting

This is an analytical technique used to detect whether the transgenic plants produce the specific protein product of the introduced gene. Protein samples are extracted from the transgenic plants, processed into denatured proteins and transferred to a nitrocellulose membrane. The protein is then probed or detected using the antibodies specific to the target protein.

Step 6. Backcross Breeding (If Needed)

Genetic transformation is usually conducted in elite or commercial varieties which already possess the desired agronomic traits but lack the important trait of the trans-gene. Thus, once successfully transformed and regenerated, the genetically modified plant will be easily recommended for commercialization if it shows stability in several generations and upon successfully passing and fulfilling varietal evaluation, bio-safety and release requirements. However, some plant transformations may have been performed in plant varieties which are amenable to genetic transformation but are neither adapted nor important in the target country. There may also be sterility problems in the transgenic plant. In such cases, conventional plant breeding is performed where the transgenic plant becomes the pollen source in the breeding programme and the elite lines or commercial varieties as the recurrent parent. Backcross breeding enables the combination of the desired traits of the recurrent parent and the transgenic line in the off-springs. The length of time in developing transgenic plant depends upon the gene, crop species, available resources and regulatory approval. It varies from 6 to 15 years before a new transgenic plant or hybrid is ready for commercial release.

Global Status of Commercialized GM Crops-2018

The first wave of GM crops was marketed mainly to farmers, with the goal of making their jobs easier, more productive and more profitable. In 1996, for example, biotechnology firm Monsanto of St Louis, Missouri, introduced the first of its popular 'Roundup Ready' products: a soya bean equipped with a bacterial gene that allows it to tolerate a Monsanto-made glyphosate herbicide known as Roundup. This meant that farmers could kill off the majority of weeds with one herbicide rather than several, without damaging the crop. Other GM crops soon followed, including Monsanto's *Bt* cotton: a plant modified to produce a bacterial toxin that discourages destructive bollworms and cuts down on the need for pesticides. Since 1996, the adoption of commercialized GM crops globally has moved fast with unprecedented scale. The latest available acreage for such crops is available for 2018 and the same is described below.

According to a detailed compilation by ISAAA (2018), high adoption of biotech crops continued in 2018 with 191.7 million ha globally. The adoption rate of top five biotech crop growing reached close to saturation with USA at 93.3% (average for soybeans, maize and canola), Brazil (93%), Argentina (almost 100%), Canada (92.5%), and India (95%). A total of 70 countries adopted GM crops in 2018 where 26 countries planted GM crops and 44 additional countries imported GM crops for food, feed and processing. Biotech crops have expanded beyond the big four (maize, soybeans, cotton and canola) and included crops like alfalfa, sugar beets, papaya, squash, eggplant, potatoes and apples all of which are already in the market. Brazil planted the first insect resistant sugar cane, Indonesia the first drought tolerant sugarcane; and Australia planted first high oleic acid safflower for R & D and seed production. Global area of GM crops by countries in 2018 is as given in Table 11.1 (ISAAA, 2018).

Table 11.1: Global area of biotech crops in 2018: by country (million hectares)**

Rank	Country	Area (million ha)	GM crops commercialized
1	USA*	75.0	Maize, soybeans, cotton, canola, sugar beets, alfalfa, papaya, squash, potatoes, apples
2	Brazil*	51.3	Soybeans, maize, cotton, sugarcane
3	Argentina*	23.9	Soybeans, maize, cotton
4	Canada*	12.7	Canola, maize, soybeans, sugar beets, alfalfa, potatoes
5	India*	11.6	Cotton
6	Paraguay*	3.8	Soybeans, maize, cotton
7	China*	2.9	Cotton, papaya
8	Pakistan*	2.8	Cotton
9	South Africa*	2.7	Maize, soybeans, cotton
10	Uruguay*	1.3	Soybeans, maize
11	Bolivia*	1.3	Soybeans
12	Australia*	0.8	Cotton, canola
13	Philippines*	0.6	Maize
14	Myanmar*	0.3	Cotton
15	Sudan*	0.2	Cotton
16	Mexico*	0.2	Cotton
17	Spain*	0.1	Maize
18	Colombia*	0.1	Cotton, maize
19	Vietnam	<0.1	Maize
20	Honduras	<0.1	Maize
21	Chile	<0.1	Maize, soybeans, canola
22	Portugal	<0.1	Maize
23	Bangladesh	<0.1	Eggplant
24	Costa Rica	<0.1	Cotton, soybeans
25	Indonesia	<0.1	Sugarcane
26	eSwatini	<0.1	Cotton
	Total	191.7	

* 18 biotech mega-countries growing 50,000 hectares, or more, of biotech crops

** Rounded off to the nearest hundred thousand.

In addition to the information in Table 11.1, as per Crop Biotech Update, 14 October, 2020), Argentina has become the first country in the world to approve drought tolerant HB4 wheat. Argentina's Ministry of Agriculture has granted approval of Bioceres Crop Solutions' HB4 wheat event for growing and consumption. The HB4 trait increases wheat yields by up to 20% and is currently the only drought tolerance technology for wheat and soybean crops in the world. Argentina is Latin America's largest wheat producer and the world's first country to adopt HB4 drought tolerance technology for wheat.

Argentina's regulatory clearance follows the approval of HB4 soybean which has been approved in the United States and Brazil. Commercialization of HB4 wheat in Argentina is contingent upon import approval in Brazil which purchases just over 85% of its wheat from Argentina. Currently, regulatory processes for HB4 wheat are advancing in US, Uruguay, Paraguay and Bolivia. Bioceres also intends to initiate regulatory processes in Australia and Russia as well as certain in Asia and Africa. Drought tolerant technology is a patented seed technology developed by Trigall Genetics, Bioceres' joint venture with Florimond Desprez, a global leader in wheat genetics

"GM technology has contributed to all facets of food security. By increasing yields and reducing losses, it contributed to food availability for more families. By enabling farmers to improve their processes and join the modern supply chain, it improved physical access to food. Through raising farmer and rural incomes, it improved economic access to food. Through rigorous standards of food safety and hygiene programmes, it contributed to better food utilization," said Dr. Paul S. Teng, ISAAA Board Chair. "While agricultural biotechnology is not the only key in enhancing global food security, it is an important scientific tool in the multi-disciplinary toolkit." Biotech crop plantings have increased ~113-fold since 1996, with an accumulated area of 2.5 billion hectares, showing that biotechnology is the fastest adopted crop technology in the world. In countries with long years of high adoption, particularly the USA, Brazil, Argentina, Canada, and India, adoption rates of major crops are at levels close to 100%, indicating that farmers favor this crop technology over the conventional varieties. More farmers' and consumers' needs, more diverse biotech crops with various traits became available in the market in 2018. These biotech crops include potatoes with non-bruising, non-browning, reduced acrylamide and late blight resistant traits; insect resistant and drought tolerant sugarcane; non-browning apples; and high oleic acid canola and safflower (press release by ISAAA on 22 August 2019).

The ISAAA report-2018 (ISAAA, 2018) also highlighted the following key findings:

- The top 5 countries with the largest area of biotech crops planted (USA, Brazil, Argentina, Canada, and India) collectively occupied 91% of the global biotech crop area.
- Biotech soybeans reached the highest adoption worldwide, covering 50% of the global biotech crop area.
- The area of biotech crops with stacked traits continued to increase and occupied 42% of the global biotech area.

- Farmers in 10 Latin American countries planted 79.4 million hectares of biotech crops.
- Nine countries in Asia and the Pacific planted 19.13 million hectares of biotech crops.
- In Asia, Indonesia planted for the first time a drought tolerant sugarcane developed through a public (University of Jember) and private (Ajinomoto Ltd.) partnership.
- The Kingdom of eSwatini (formerly Swaziland) joined South Africa and Sudan in planting biotech crops in Africa, with the introduction of IR cotton. Nigeria, Ethiopia, Kenya and Malawi granted approvals for planting IR cotton opening Africa to biotech crop adoption.
- In Europe, Spain and Portugal continued to adopt biotech maize to control European corn borer.
- More area planted to biotech crops for farmer and consumer needs included potatoes with non-bruising, non-browning, reduced acrylamide and late blight resistant traits; non-browning apples; insect resistant eggplant; and low lignin alfalfa, among others.
- New crops and trait combinations in farmer fields include insect resistant and drought tolerant sugarcane; high oleic acid canola and safflower.
- Various food, feed and processing approvals for Golden Rice, *Bt* rice, herbicide tolerant cotton, low gossypol cotton, among others.
- Cultivation approvals for planting in 2019 include new generation herbicide tolerant cotton and soybean, low gossypol cotton, RR and low lignin alfalfa, omega-3 canola, and IR cowpea, among others.